Paradoxical Life

PARADOXICAL LIFE

Meaning, Matter, and the
Power of Human Choice

Andreas Wagner

Yale University Press / New Haven and London

The publisher gratefully acknowledges permission to print
a translation of Paul Éluard, "Le Droit le devoir de vivre,"
in *Le Livre ouvert II: (1939–1941)* (Paris: Gallimard, 1947)
Editions Gallimard, Paris.

Set in Galliard type by Keystone Typesetting, Inc.,
Orwigsburg, Pennsylvania.
Printed in the United States of America.

Library of Congress Cataloging-in-Publication Data
Wagner, Andreas.
The inner dialogue of creation : how paradox creates a
world of meaning, choice, and possibility /
Andreas Wagner.
p.   cm.
Includes bibliographical references and index.
ISBN 978-0-300-14923-4 (cloth : alk. paper)
1. Biology — Philosophy. 2. Paradox. I. Title.
QH331.W285   2009
570.1 — dc22
2009004609

A catalogue record for this book is available from the
British Library.

This paper meets the requirements of ANSI/NISO
Z39.48-1992 (Permanence of Paper).

10 9 8 7 6 5 4 3 2 1

Men will never fly,

because flying is reserved for angels.

Milton Wright, father of the Wright brothers, 1903

. . .

Our deepest fear is not that we are inadequate.

Our deepest fear is that we are powerful

beyond measure.

Nelson Mandela, 1994 inaugural speech

CONTENTS

CONTENTS

This book weaves together recent insights from almost all areas of modern biology to make two central points: fundamental paradoxical tensions permeate the biological and nonbiological world; these tensions give humans great power and a central role in creating this world. This book is written not for an academic expert but for a general reader who is interested in our place in the world and the implications of recent scientific insights for this place. The facts described in this book are not new, but their juxtaposition is. The expert will find many simplifications. I discuss some of the important nuances of these simplifications in notes, but I am acutely aware that the depth of treatment might not satisfy academic philosophers or biologists. My goal has been to outline a complex argument in a simple and readable way and to leave an exhaustive tome to a future publication. A desire for brevity also explains some of the more glaring omissions, such as the absence of immunology from a discussion of the tension between self and other.

I am very grateful to Betsy James, who has gone over the manuscript several times. Many thanks also to Jean Thomson Black, Luis Cadavid, Hilary Hinzmann, Jonathan Kaplan, and Günter Wagner, as well as two anonymous reviewers, who have given earlier versions of the manuscript a thorough reading and provided many valuable comments. Numerous colleagues provided critiques of individual bits and pieces over a long period of time. This book would not have been possible without them.

Paradox and the Power of Choice

We will either find a way, or make one.

Hannibal, Carthaginian general, before crossing the
Alps 218 BC with an army of fifty thousand

If the world were perfect, it wouldn't be.

Yogi Berra

In the womb, all human body parts, be they simple like hair and nails or intricate like eyes, form through an unimaginably complex communication process. The participants in this process are millions of cells. These cells release thousands of molecules as signals that crisscross the embryo and contain instructions to other cells: move over here, move over there, bulge, flatten, swell, shrink, divide, or die. The respondents follow their instructions and talk back. This cellular conversation is as carefully orchestrated as a symphony but vastly more complex. The result is a human body, including the very eyes that read these lines.

The making of an embryo, the attack of a deadly virus, the building of a termite's mound, and human conversations all require communication, which rests on meaning. Whereas our language conveys meaning through words, other languages speak through body movements, odors, molecules, or electrons. In other words, the material carrier of meaning differs among communication processes. But nonetheless, almost any process in the natural world communicates meaning from one thing or being to another. Not just that, all matter is potentially endowed with meaning—to something or to somebody.

Meaning is not secondary to matter, as much of twentieth-century science would have us believe. We could instead view all matter as carrying meaning, and all meaning, even the most fleeting thought, as having a material aspect. Like the two sides of a coin, meaning and matter are completely separate—opposites even—and completely inseparable.

Paradoxical relations like this and the following are central to this book and to the world.

To defend their home, glue-grenade ants detonate themselves by violently bursting a gland, thus entrapping any intruder in an immobilizing glue. In a cancer, cells divide selfishly, sucking resources from the body that hosts them; their greed as they destroy the body eventually destroys them, too. A nation at the brink of nuclear war may concede to its opponent, thus avoiding their mutual destruction. It acts out of self-interest but ultimately also serves its opponent. These three examples point to another important tension, the tension between self and other.

Billions of years ago, all of life consisted of single cells. Rapid reproduction and efficient survival were the name of the game. Cells successful at both endured. All others perished. Because life was all that mattered, competition for it was undoubtedly ruthless. But at some point, hidden in the recesses of deep time, the rules changed. Cells not only abandoned ruthless competition but also learned to cooperate. Not only that, they learned to sacrifice their own lives. Just take our bodies. Billions of body cells, in our brain, blood, muscles, and bones, will die a certain death. Only a minuscule fraction of sperm and egg cells are potentially immortal. How did life come to be that way? How did cells, unthinking lumps of protoplasm, learn to sacrifice life itself?

Many biologists think that no living thing ever acts truly unselfishly. According to this view, altruism—even in the extreme form of the glue-grenade ant or your body's cells—is selfishness in disguise. But others feel that this perspective cannot be the whole truth, that most of what we call "good" originates in true altruism. I offer a third perspective: self and other are utterly different and separate, but where the action of one — cell, organism, or nation — influences the fate of the other, self and other are also inextricably linked. Again, like the two sides of a coin, self and other are utterly separate yet inseparable.

These are just two examples of paradoxical tensions that play an important role in this book. Others include the tension between creation and destruction (one always accompanies the other, whether in the making of an organism or the building of a nation), the tension between part and whole (are your genes, you, or the world around you responsible for your actions?), or the tension between risk and safety (organisms play deadly games with their genes, games, however, that have allowed them to survive).

In a paradox, a statement and its opposite are inevitably true or inevitably false. A statement such as "Meaning and matter are completely separate" is of this nature. You might think that paradoxes are just mind games and that they live "just" in language. That would be an understandable impulse, given how paradoxes confuse and bewilder us and how they render the world much less orderly than we would like. It would be understandable to insist that any paradox must have one right resolution or to dismiss it completely. But early in the twentieth century, mathematicians began to discover paradoxes that cannot be ignored. These paradoxes occur in the very foundation of mathematics, on which natural science rests. One can even build machines, or computers, that embody these paradoxes. These paradoxes are thus far from mere games, and they reach much further than human language. The paradoxical tensions at which I have just hinted occupy much the same place: they are built into the foundation of the world. They are everywhere.

Why care about all this? *Because to accept paradox as fundamental gives humans enormous power.* Three hundred years of Western science have deposed us from the center of the world and closer to its fringes. We have become gypsies at the edge of the universe, as the biologist Jacques Monod put it.[1] The power of paradox, however, reinstates humans in the world's center.

To accept fundamental paradoxes means the end of absolute and final knowledge. It also means that we often have to choose sides in a paradox — whether ourselves or others are primary, nature or nurture, matter or mind, and so on. To understand, we must choose to accept truths knowing that any such truth, taken to its final consequence, will turn out to be false and thus limit our understanding. The power of such choice is exquisitely human, and it gives us an important part in the creation of a world. The world itself is inextricably linked to our choices. They are choices of how we view the world—in mundane everyday life or in sophisticated scientific theories—and choices of how we shape it by our actions.

Consciously or unconsciously, we make such choices all the time. The wisdom of all human cultures, as enshrined in great art, in folktales, and in holy texts, tells us that it is difficult to live fully aware of our choices, instant by instant, but that it is essential to cultivate that awareness if we hope to fulfill our human potential. To attain that awareness is the examined life of Socrates, Plato, and Aristotle, the satori of Buddhism, and the "peace that passeth all understanding."[2] Commitment to choice opens many doors and carries tremendous power.

Much of this book is a journey through biology. Why biology? Paradoxes have always been of central importance to logic and mathematics. Some famous paradoxes, such as the liar's paradox, have challenged scholars for centuries. (The statement "This sentence is false" encapsulates the liar's paradox. Is this statement true or false?) In contrast, paradoxes in biology are less familiar, though no less plentiful and no less challenging.

Biological paradoxes also have a great advantage over more abstract logical and mathematical paradoxes: we can observe them in action everywhere, from humans down to the humblest bacterial cells and the molecules within them. More than anything, biology's paradoxes illustrate the great unity of life. Many of the most difficult problems and conflicts humans face are shared by the lowliest bacterium. Take the tension between self and other, which governs every conflict there ever was, from geopolitical decisions that shape national boundaries and the fates of millions to true arms races among single-celled organisms: self and other play no role in the world of mathematics. They were born with the living.

Because the world is rife with paradoxes and because one can get hopelessly lost in any one of them, I propose two rules to deal with paradox.

First, when encountering a paradox, here or elsewhere, do not dismiss it. It has been said that paradox is like a gatekeeper to knowledge, denying passage to those who avoid it. Therefore, explore the paradox, examine all its strange facets, plunge into it.

Second, once you are in danger of becoming hopelessly entangled — a danger as real as paradoxes — choose one side and stick with it to see where this choice takes you. But never forget that you made a choice. The more we stretch our ability to choose — here or anywhere — the more we gain. And the more we have to give up. If we are bold enough, the world as we know it may be dissolving before our eyes. This prospect is terrifying. It also — paradoxically — gives us immense freedom to create.

Once you can stand in a paradox — not that it will ever be comfortable — you may be able to glimpse beyond it. Such glimpses have been the goal of every great philosopher, scientist, mystic, and theologian. And this "something" behind paradox is what should be the book's real subject, though I have not succeeded in writing about it. To do this subject justice might take an endless conversation or one that does not take place at all — like a circle with an infinite diameter or a perfect mathematical point. Thousands of years ago, the Chinese sage Lao-tzu had already expressed it best when he said, "Heaven and earth begin in the unnamed."[3] Here, I can only point to the something behind paradox. The rest is up to you. As those report who have seen it, a mere glimpse will reward you beyond imagination.

CHAPTER 1

The Inner Dialogue
of Creation

The universe is made of stories,

not of atoms.

Muriel Rukeyser

Everything is a metaphor.

Johann Wolfgang von Goethe

The kind of communication most familiar to us is a conversation between two people. We tend to think that human conversation is vastly different from other forms of communication, not least because it exists only in our species. In so doing, however, we overlook that animals, plants, bacteria, and even molecules carry on conversations that may exceed our own in subtlety, flexibility, and the power to shape the world.

In writing lines such as these, a writer experiences a continual inner dialogue. For example, I ask myself, was this last sentence a good way to get my point across? Then I make a decision. Yes, I am going to let it stand as is for now. What do I want to say next? What I want to say is that this dialogue with oneself, examining one's standpoint, making a choice, and then moving on to the next sentence, is going on in our heads all the time. So let's just say that. What next? I could work in a sentence about the chattering neurons that are important for this dialogue. But maybe I should save that for later? Yes, let's not talk about neurons just yet.

Every book is also a subtle conversation between writer and reader. In the hope of making this book satisfying and enjoyable to you, I imagine how you are going to perceive what I say. And every moment, I make a choice based on what I think you might think. Although you are not even here, I am having a kind of conversation with you.[1]

As soon as this conversation of writing has ended, I will communicate with a publishing house, with the intent of having a book created. This communication begins with a letter summarizing for whom I have written the book and what I hope it will accomplish. If the publisher decides to publish the book, an editor will converse with many others involved in its editing, production, sales, and distribution, including a manuscript editor, a designer, a production manager, publicists, marketing and sales people, distributors, and booksellers. Will the book have an index? What are the best title and subtitle? What size will it be? What will make an attractive cover design? What kind of paper should be used? These conversations involve everybody from high-level managers to paper suppliers and production floor operators. The product of all these conversations is the book you hold in your hand.

The point is that conversations and the world at large are much more interwoven than we think. Conversations can have material consequences, such as this book. Some consequences can change the course of history. Just consider Karl Marx's *Das Kapital* or the Bible and how they have changed the world.[2]

What begins as an author's inner dialogue can thus lead to material changes in the world. Does inner dialogue occur only sometimes and in some people? No. It rambles along all the time as a ceaseless, inaudible chatter in your head. (Have you just asked yourself: "What is he talking about?" There it is!) Let's listen to one man's inner dialogue as he might prepare for a job interview: "What should I wear? The blue suit. Now what about this green tie? No, too shrill. The black shoes? Nope—too worn. And the gray overcoat? Conservative, but that's okay for an interview."

This interviewee is concerned about a facet of human communication that we have not yet encountered. He is concerned about sending the wrong message to the interviewer—not in words but with a different signal: the clothing he wears. Far from just protecting a delicate human body from the elements, clothing may, in fact, be more important in communication than in protection. Not all cultures use clothing for protection against the elements. But all cultures use body decoration to communicate.[3] Clothing, jewelry, and perfumes convey a stream of messages about social standing, emotion, profession, religion, sexuality, ethnicity, and cleanliness. And bodily adornment is only one of many ways people communicate without words. The interviewee's facial expressions, postures, mannerisms, dialect or accent, address, previous employment — all these tell a story to the interviewer. And in doing so they become a part of the job interview, affecting the kinds of questions asked and, in turn, their answers.

Most human encounters involve communication. And communication can be as varied as the people involved. In some kinds of communication, words are central. In others — flirting, dancing, lovemaking — words may play a lesser role or none at all. Some communication, wordless or not, involves two individuals. Other kinds of communication involve many people. (Think of an orchestra performing a symphony.) One can thus conceive of endlessly many kinds of communication.[4] A few distinctions among them are important for us, distinctions to which I now turn.

A World of Signs

The elementary particles of communication are signs, which convey meaning.[5] We often do not appreciate that signs can have any conceivable material form, such as sheets of painted tin (mileage markers on a roadside), molecules drifting through the air (the smell of food), or water vapor growing into immense castles in the sky (announcing a brewing storm). Is there any feature that unifies all signs, any feature they have in common? They all stand for something else. The cross on the necklace of a Catholic woman stands for her religious affiliation. The interviewee's drumming fingers stand for his nervousness. And a putrid smell stands for a rotting carcass. We could, of course, spend endless hours finding more elaborate distinctions among signs. But for our purpose this will suffice: "A sign is any object, state, or event used to indicate another object, state, or event."[6] It is also useful to have a name for what a sign stands for, points to, or indicates. Let's just call that its meaning.[7] Philosophers have spent more than a century trying to pin down the meaning of "meaning" in more precise terms.[8] Despite their efforts, the matter is still unsettled. Consider that the reason for this failure is not incompetent philosophers but that it may lie deep in the subject matter itself. Meaning is as central to the world at large as matter is to the world of physics, which prevents us from defining "meaning" in more elementary terms. Meaning as "what a sign stands for" will have to suffice for our conversation.

Signs that resemble what they stand for are called icons. Iconic signs are images of what they point to. They share some inner similarity with it. A simple example is a photograph or a sketch of a landscape, person, or building. Others include metaphors and scientific theories.[9] Signs that lack such inner similarity are called symbols.[10] Symbols carry meaning by an agreement among those who use them. Human language signs like "this book" are an example.[11] Hand signs, such as the secret sign language of Freemasons or the signs used by airport ground personnel to guide a landed plane, are others. Both symbols and icons are involved in every human conversation, including the job interview described earlier. The symbols in that interview include spoken words and written résumés; its icons may include the company's performance chart; and yet other signs, for example, the interviewee's conservative overcoat, might occupy an in-between status.

The example of the interview also shows that intent, wishing to send a message, is not needed for communication.[12] To be sure, the participants of any interview choose many of the signals they send, such as the words they utter, the clothes they wear, and how firmly they shake hands. But they may not choose many other signals, for example, their gestures, posture, and facial expressions. Some signals, even though they convey crucial information, may be impossible to choose, such as skin complexion, gender, or height. Tall managers, for example, are promoted more often because they are perceived as more impressive.[13]

So much for communication among humans. Nonhuman communication, as we will see, parallels human communication in many ways.[14]

Plants and Parasites

Most of us can easily name examples of communication among animals: the neon colors of poison arrow frogs in the Amazon rain forest warn predators to stay away; the male peacock flashes brilliant plumage in hopes of attracting females to mate; and the gaping mouths of baby birds mean one thing to their parents: feed me. These rather obvious examples involve animals that are similar to us in one respect: they have a complicated nervous system similar to ours. But many striking examples of communication do not involve animals like these. Some do not involve animals at all.

Many insects make their living on or inside plants. Some reproduce on their host plants, laying eggs on a surface such as the underside of leaves. The eggs develop there and eventually hatch into leaf-eating larvae in a paradise of plentiful food. Other insects bury their eggs deep inside part of the plant, such as a ripening fruit or a tree trunk. The hatching larvae then feed inside (and on) the plant's body.

Insect parents must be judicious in picking the plant that will host their offspring.[15] Just think what the consequences would be if a hundred female insects chose the same plant for their eggs. The hatching larvae would face a life-and-death competition for food, and many would perish. An egg-laying female needs to avoid this competition. She must avoid plants where other insects have earlier laid eggs. And after she has found a free host plant, her eggs should deter other females from using that plant. Insects have mastered all of these skills. Some mark the location of their eggs with a substance exuded by their abdomen. Other females, when they detect this substance, will stay away from the plant. In some species, the developing eggs exude a chemical signal announcing their presence to other insects. In yet other cases, insects manipulate the plant into producing a substance that reveals their presence; here the plant—unbeknown to itself—provides a sign that stands for the presence of eggs.

The result of all this messaging is a peculiar conversation among female insects looking to place their offspring on the perfect host plant. This conversation is highly complex: imagine hundreds of insects trying simultaneously to find the best among many host plants. Each responds to signs provided by other females and sets signs for them in turn. Often she can detect not only that others were there before her but how many eggs were laid, even though they are concealed deep inside a ripening fruit. This information can help her decide whether she should lay more eggs in the fruit or look elsewhere.

Insects are not the sole participants in this intricate conversation. What about the plants themselves? They surely have no interest in being devoured by starving larvae. Not surprisingly, many plants defend themselves against the feeding insect hordes. Some rely on communication in this defense. When damaged by feeding insect larvae, these plants produce a chemical signal, an alarm bell. This signal alerts other insects, the natural enemies of the plant's invaders, to come to the rescue. These insect enemies devour the plant-eating insects or their offspring. To these enemies, the chemical signal means that prey is near. It means "food."

In yet another layer of this complex conversation, those female plant-eating insects that are still looking for good host plants can also hear this alarm bell and may not lay their eggs on the plant. To them, the chemical signal means two things: first, others have already claimed the host plant; second, as a result, the host plant is now very appealing to enemy insects.[16]

The natural enemies of plant-eating insects add yet another facet to this conversation. Some of them—called parasitoids—have a peculiar lifestyle. Their females lay eggs inside the growing eggs or larvae of plant-eating insects. As the plant-eating insect's eggs grow or as their larvae crawl on the host plant, the parasite's eggs grow inside them and hatch into larvae themselves. These new larvae then devour the plant-eating larvae from the inside out killing the host larvae in gruesome fashion.

The natural enemies of plant-eating insects also need to know whether another egg-laying female has already visited the host plant. That is, without being able to see and count the eggs inside an egg, they must know whether their host egg already harbors other eggs. Fortunately, just as plant-eating insects mark the plants they have invaded, their own enemies mark the insect eggs they have invaded with an invisible ink. Some mark the outside of the host egg. In this case, a newly arriving female can easily find the message. Others mark the egg's inside. In that case, a newly arriving parasitoid must probe the egg to find out how many parasitoids are already in it. There may or may not be room for more.

And the conversation does not end here. The parasitoids themselves have natural enemies, some of which are also parasitoids and likewise lay eggs into the eggs of other parasitoids. And a dialogue similar to those above ensues. Which eggs already hide eggs? How many do they hide? Is it better to move on to another host? Further, the parasitoids' enemies may themselves have enemies that are parasitoids, and so on, a nested set of enemies that feed off one another.

Many Parallels to Human Communication

In this multilayered communication network plants deter insects and attract their enemies; insects of the same species send signals to one another to avoid competition; and insects of one species reveal information — inadvertently — to the parasitoids that will spell their doom. To be sure, this form of communication uses a different — chemical — language than the conversations you and I are most familiar with. But it is by no means less complex. And it contains many elements we are familiar with from human conversations.

First, this communication network uses signs that point to something else, such as molecules that point to the presence of a plant-eater or to competing eggs or larvae. Second, these signs are often not sent on purpose: think of the signal an egg-laying female sends — unwittingly — to its natural enemy.[17] Third, the conversation carries many opportunities for misunderstanding. For instance, some egg-laying mothers mark their host fruit with excrement. That excrement is supposed to deter other mothers from laying eggs into the fruit. But if another mother encounters excrement on a fruit, she does not know whether a male or a female left that excrement. Males do not lay eggs. Thus, in a profound misunderstanding, male excrement can deter mothers from egg laying just as female excrement would.[18] Fourth, like humans, insects may attempt to avoid such misunderstanding, such as by sending the same message twice. For example, some insects mark the interior and exterior of their host, perhaps in an effort to ensure that newly arriving females get the message. Fifth, eavesdropping — often to the detriment of a message's sender — is rampant. Just take the parasitoids that home in on a female's mark intended for other females. To these females the mark means, "This plant part is taken." But to a parasitoid it means, "Here is a host in which to lay my eggs." This example also illustrates that a signal that means something to you may mean something else to me, and it may mean nothing at all to a third party. And this conversation has many layers with blurry boundaries. That is, each layer is linked with others. Some parasitoids exploit a plant-eating insect's mark and others a plant's chemical alarm bell, but they all link their conversation to those between plant-eating insects and host plants.

Conversations among Many

In bumblebees, males explore the world along certain flight paths.[19] That is, any one male will stick to his particular flight route, like a truck driver preferring certain highways to cross the country. Along their route, flying males mark objects with a chemical—a scent—from their labial glands. Other males are attracted to these marks. Through this mutual attraction, multiple males share parts of their flight path. (In some tropical bumblebees, more than five hundred males may share the same path, each marking objects along the path.) But not only males find these scents attractive. Females do, too. The more marks an object has, the more likely a female is to land there. Having landed, females may initiate a communication process with passing males that will eventually lead to mating. Put differently, many males must have landed and marked a site before a female will land and initiate mating. There is little point in looking at this form of communication only from the viewpoint of two participants—perhaps one male and one female—if so many others are necessary to initiate it.

Social insects like bumblebees, ants, and bees provide many other examples of extremely complex communication processes involving thousands of participants. The results of such communication rival the most astounding feats of human technology. Through communication, insects can determine the distance to a food source, the quality of the food, or the volume of a cave suitable for nesting. They can also build extremely complex colonies with sophisticated air-conditioning systems, cities home to thousands of individuals.

One could argue that insect communication—involving organisms we like to think of as simple—is mechanical. That is, its participants just follow some program hardwired into their tiny brains. The examples above already hint at how limiting this view is, because in many of them, learning is a key element. And there are many others where the participants adapt equally flexibly to a situation, such as communication between ants and some lycaenid caterpillars.[20] The Lycaenidae are a large family of butterflies with some six thousand species. In the caterpillar stage, Lycaenidae have a gland on their back that produces sweet nectar. They lure ants to this gland by vibrating their body and through chemical messages. When an ant arrives, a caterpillar feeds it sweet nectar. The caterpillar is generous, because ants defend it against natural enemies such as parasitoids.

How vigorously a caterpillar courts ants and how much nectar it hands out depends on the degree of danger it senses. This degree of danger, in turn, depends on the number of nearby predators and on the number of nearby caterpillars. An ant in turn will defend caterpillars more aggressively when it is awarded more nectar. Even plants have such reward systems: many plants depend on insects to pollinate and fertilize them. In return, they provide insects with nectar. Plants can learn whether their visitors pollinate effectively and increase or reduce their nectar reward correspondingly.[21]

In all these conversations, none of the participants need to understand — in any human sense — what the signals mean. (Some of them, like plants, do not even have a brain or a nervous system.) In addition, the participants may not know whether they are conversing with one or more others, a feature that has parallels in humans. In humans, a conversation between two people never strictly involves solely them. A conversation between two people depends on previous conversations or communication each participant had with many others, and on communication among others apart from the two participants. The appropriate thing to do, say, think, or feel in a conversation influences each person's sense of what to communicate and how to do so. The most private communication is, from this perspective, a minute part of a much larger human conversation.

Microscopic Conversations

Communication may occur between individuals or the same species and between species; it may occur between animals and plants; and it may involve microscopic organisms, such as plankton. The planktonic world contains innumerable small animals, some with many cells and even a tiny nervous system, others with only one cell. It also contains many plants, mostly floating algae ranging again from one-celled organisms to intricate laceworks of green tissue. Like the macroscopic world of humans, this microcosm contains countless stories of cooperation and exploitation, prosperity and famine, creation and destruction.[22]

As in our world, inhabitants of this microscopic world have to make a living. This need creates numerous conversations among friends, foes, and friends and foes. The smaller a planktonic animal is, the greater is its danger of being eaten by other animals. For some this fate lurks in the form of predators that like to hide in tangles of planktonic plants, microscopic floating jungles that provide ideal shelters for predators. To their prey, the dark tangles mean something—looming danger—and they stay away from them. Other prey animals respond to chemical signals released directly but unwittingly by a predator. You might say they respond to the stench of destiny. Some prey respond to—interpret—this message in an astonishing fashion: they transform themselves into floating fortresses, studded with protective plates of armor or bristling with sharp spines or teeth. They literally change shape and grow new defensive body parts.

The resulting transformation can be so radical that even experts might not recognize the animal afterward. A floating fortress like this sends a clear message to a predator, similar to the message of a forbidding castle to an attacking army. But this transformation is costly. Its price is similar to that of fortifying the walls of a castle and maintaining increased vigilance, a cost to be paid only when absolutely necessary. If there are few or no predators around, the prey animal changes back to its less defensive shape. Over the course of several seasons, animals change back and forth repeatedly, depending on their sense of danger. In this complex communication between predators and prey, prey detect the presence of predators, perhaps even their numbers. They react accordingly by putting on armor. And predators react to the messages sent by its bristling prey and do not pursue them. As the number of predators and the ensuing risk falls, prey can shed their armor—until they detect another wave of predators creeping up on them.[23]

Such protective communication would often fail if it involved only two individuals. Many animals, for example microscopic water fleas, form swarms when predators are near.[24] Their predators cannot single out one large and therefore desirable animal in a swirling mass of prey. They will thus often blindly lash out at the swarm. Such a haphazard hit-or-miss attack has a much smaller chance of success than the targeted pursuit of one animal. In other words, swarms provide protection in numbers. But animals pay a price for swarming. When hundreds or thousands of organisms are huddled in one place, they may run quickly out of food and so risk starvation. It thus makes sense to swarm mostly if predators are around. Tiny planktonic animals can react to the presence of predators, probably by reading the message their scent leaves in the water. Each prey animal sends a signal to others of its kind indicating its presence and location. In turn, it detects signals sent by these others and moves toward the loudest signal source. When thousands of animals do the same thing, the result is the throng of organisms we call a swarm. Once the air is clear, the swarm can disperse.

Communication also exists on an even smaller scale, involving organisms such as single-celled bacteria. They do not have a nervous system, yet they can communicate using elaborate molecular signals. A fascinating example is the marine bacterium *Vibrio fischeri*, a peculiar light-producing creature that occurs in enormous colonies inside a squid host.[25] Within the squid, many *Vibrio fischeri* cells cram together in a small, specialized organ, where they produce light that the squid uses to hunt for prey. As a nocturnal predator, the squid uses the bacteria to illuminate the ocean floor below, and this light also neutralizes the shadow of its body cast by the moonlight. The squid thus avoids signaling its presence to predators. The bacteria in turn receive food from the host. These bacteria light up only when millions of them are jammed into a small space. Although each bacterium releases a chemical signal that the other bacteria can detect, the signal quickly dissipates in the open ocean. But millions of cells each releasing and detecting this signal produce so much of the signaling chemical that they can determine how many others are nearby. If each cell detects only a faint whisper, it will not light up. But if that whisper turns into a deafening roar, it will. And so will all the other cells around it.

Communication among bacteria is the rule rather than the exception. Bacteria in general communicate. For example, many cause infectious diseases, but one bacterium is not sufficient to cause disease. A successful infection needs hundreds or thousands of bacteria. And a bacterium may not even start the process of infection until it senses many others of its kind around. Some bacteria can form very tight associations called biofilms. Such biofilms serve to protect the individual bacterium against a hostile environment, and they allow the bacteria to exploit this environment. One example of a biofilm is the bacterial plaque that forms on your teeth when you do not brush them. This plaque is hard to remove, because the bacteria in it hang together. They protect themselves against their prime enemy: you and your toothbrush. To do so, they produce a kind of glue that bonds them. Obviously, there would be no point for a lone bacterium to produce this glue. Thus, before ramping up glue production, bacteria communicate to detect how many others are present. If there are enough of them, they start building their biofilm. Their communication continues throughout the making of the biofilm, in which different cells take on specialized roles. And as in any human society, the bacteria's division of labor requires extensive communication.

All these forms of communication have many parallels to human conversation.[26] Some bacteria communicate only with their own kind, whereas others — you might call them multilingual — communicate with other kinds of bacteria.[27] The sender of a sign may not intend to send it: recall the tangle of floating algae and the message it conveys to a tiny animal in the plankton. The meaning of a sign depends not only on the sender but also on the receiver: a male scent mark meaning the world to a female bumblebee may mean nothing to a passing wasp.

Matter and Meaning

All communication contains an inextricable, paradoxical link between matter and meaning, which are as two sides of a coin, opposites yet completely inseparable. When thinking about human conversation, we usually think about the information exchanged or about the concepts that are at a conversation's heart. And because we converse effortlessly, with little regard to how we converse, we act as if these concepts had a life of their own, as if they could exist by themselves. We often forget that any human conversation requires a material carrier, be it light for visual signals, air molecules and their compression in sound waves, or chemical signals like smell and taste. In fact, human conversation usually involves several carriers. But most important is that any conversation, from that inside our head — mediated by ions traveling through nerve membranes — to that of bacterial cells gauging their numbers, needs matter. And because matter is far removed from what we consider the heart of a conversation, we tend to overlook its importance. We overlook that there is no conversation, no meaning without matter, and similarly, that any change in the meaning of a signal requires a change in matter. Matter impacts meaning.

This coin also has an opposite side, often equally neglected: meaning impacts matter. I have discussed complex human conversations and how the meaning they create can change the world, including matter in it. This principle is also illustrated by some of the aforementioned non-human communication, such as the planktonic organisms that interpret a predator's chemical signals. Their interpretation is a transformation of themselves — of whole animals — into a swimming fortress. This transformation of matter occurs in response to a signal, a carrier of meaning. The following examples point more directly to this second side of the coin: the impact meaning has on matter.[28]

Nascent Conversations

The most profound and magical transformation of matter — the creation of an organism — reveals how essential meaning is to transform matter. After conception, you and I spent about nine months in our mother's womb, developing from single cells into complete organisms. During this time, each part of our bodies was formed: lungs, heart, limbs, brain, and the eyes we use to read these lines.

Eye development illustrates strikingly how organisms and their organs are created. This process involves an extremely complex form of communication, similar to the communication among bacterial cells, except that the cells are locked inside the human embryo. Eyes start to form through a conversation between two groups of cells: the neural tube and the ectoderm.[29] The neural tube is a group of cells from which the brain and the spinal cord will eventually form. (In fact, only a tiny part of the neural tube is involved in forming eyes — the part that will later become the brain.) The ectoderm forms a flat sheet surrounding the embryo (think of it as the embryo's skin).

A few weeks after conception, cells of the neural tube initiate a communication process with nearby ectoderm cells by sending them a message, a stream of molecular signals released in the thousands. These signals reach the ectoderm in a matter of seconds because the two tissues are separated by mere millimeters. The message is akin to a command. It says one thing to the ectoderm cells: thicken! In response, this flat sheet of cells thickens profusely, until it has formed a bulblike protrusion, a lump of cells pointing inward from the embryo's surface. Later, this lump will detach from the ectoderm, and much later, it will become the eye lens. But during its formation the eye lens already sends chemical signals to the neural tube. These signals instruct the neural tube to form a cuplike shape around the future lens. The neural tube's interpretation is to form a structure we call the optic cup. Together with the ectoderm and the forming lens, the optic cup plays a key role in shaping other eye parts, from the light-sensing retina to the iris to the muscles that deform the lens to focus light. And all these parts, as they are being formed, send and receive signals shaping one another. The cornea — the hard skin covering the exterior surface of your eye — forms in response to another signal of the lens that is different from the one making the neural tube form the optic cup. The lens, in turn, cannot continue to form without further signals, one of which comes from the vitreous body, a transparent mass between optic cup and lens itself. The vitreous body sends a signal to the lens that makes the lens produce the light-transparent cells that are at the heart of what it does — to focus light on the retina.

At first sight, all this may seem a rather simple form of communication, with some cells sending a handful of signals back and forth. But the few examples I mentioned cannot do justice to the complexity of this process. Think of a forming eye as a symphony of signals involving hundreds of musicians, only some of which we can distinguish in a microscope. Each individual eye part, even the shape and location of each cell in the eye, is the result of a finely orchestrated conversation complicated beyond measure. It all begins with the thickening of the ectoderm and does not cease for a moment until the eye is formed, a fortissimo of signals crisscrossing a tiny speck of tissue. And this conversation probably continues throughout your and my whole life, maintaining the eye's function.[30]

What holds for the eye holds for every other body part, the heart and its contracting chambers, the brain and its byzantine labyrinth of folds, the lungs and their fractal tree of air ducts. Each is the product of conversations complex beyond imagination, held millionfold every single moment, inside billions of organisms that form on earth during every moment.[31]

The Mutual Shaping of Lives

To build one organism, communication among its parts may not suffice. The shaping of living things may also require other living things. That is, different organisms may need to communicate with one another to take shape. Such communication occurs not only between related organisms, such as a mother and her embryo, but can even be required to form unrelated organisms whose relationship is anything but friendly.

Witchweeds are parasitic plants that make their living sucking nutrients from other plants. Their hosts include important food crops from India and Africa, such as sorghum and millet. From tiny seeds, witchweeds grow into plants that form a network of subterranean feeding tubes. This network penetrates its host's tissues and extracts nutrients from it. Hundreds of witchweeds can feed on one host plant. Overwhelmed by these lifeblood-sucking parasites, the host plant often dies. An infestation of witchweeds can destroy both crops and farmers' livelihoods.

Each witchweed produces thousands of tiny seeds that are scattered through the air. Some of them may land near a potential host, others far away. The first step in forming a witchweed, as in most plants, is germination. To germinate, however, a witchweed seed waits for a signal from the host plant. This signal might consist of just one chemical or of a complex mixture of chemicals. The plant sends this signal inadvertently (every organism releases chemicals that reveal its presence). To a witchweed seed, the signal means one thing: germinate. The seed now transforms into a tiny seedling, its shoot pushing upward toward the surface and its root pushing down into the earth. Witchweed seeds that have fallen far from a host plant cannot detect this chemical signal; it is too faint, perhaps inaudible. These seeds do not germinate, but they can remain dormant in the soil for years. After germination, a witchweed seedling has no means of supporting itself. As a seed, it had only the food store provided by its parent, which it consumed to grow its first root and shoot. Most plant seedlings would now break through the surface and use their leaves to produce energy from light. But not witchweed seedlings. After germinating, they search for chemical signals from nearby host plants. On detecting these chemical voices, they grow feeding tubes toward the host.

Having found the host's roots, the feeding tubes embrace the roots like so many tiny, many-armed wrestlers. They then begin the next step in making a witchweed. They need to penetrate the host's network of nutrient conduits. And for witchweeds and many organisms with similar lifestyles, entering the host requires reading a chemical message from the host. This message specifies where to enter the plant and may contain instructions on how to proceed after that.

So far this form of communication appears somewhat one-sided, a host monologue into which witchweeds eavesdrop and of which they take advantage. But not only are host plants crucial in molding witchweeds, but witchweeds also reorganize their hosts. And they use chemical signals to do so. For example, witchweeds influence the numbers of the host's parts, such as leaves that grow above ground. And hosts respond to the presence of witchweeds by producing more roots below ground. Many details of this chemical communication are unknown, and it is certain to be much more complicated than we appreciate. For our purposes, it is one example of how communication molds both organisms. Many other examples exist, and countless more await discovery.[32]

It may still be difficult to see these disparate examples as instances of one and the same, of a conversation. If so, we need to examine the main obstacle in making this choice of perspective: at first sight, a conversation and any ensuing change in the world seem very different. Isn't there something irreconcilable and utterly different about the signal of a predator and a water flea fortifying itself? Or between the chatter of bacteria and their transformation into living lightbulbs? Or between the thousands of messages exchanged among living tissues and their response of thickening, twisting, and contorting themselves into any conceivable shape? Do they not reflect separate worlds, one involving communication and meaning and the other matter and mechanics, like — apparently — the printing of a book or the building of a house? These questions bring us back to the paradoxical tension between meaning and matter. Like the two sides of a coin, they are indeed utterly different. That much is easy to see. Maybe less obvious is that they are also inseparably connected.[33] A careful look at the molecular details underlying these conversations may help us see this. It will show that the link between these conversations itself involves communication, communication between molecules.

Molecular Communication

An intimate link exists between the signal a water flea receives and the animal's self-transformation; the signal a lycaenid caterpillar receives and its nectar production; the signal a marine bacterium receives and its illumination; the signal the ectoderm receives and its formation of a lens; and the signal a witchweed seedling receives and its invasion of other plants. They are linked by a network of thousands of events that take place inside an organism. And although we do not know all the details, we know that these events, the links between signal and response, are very similar among vastly different forms of communication. I can thus focus here on one example, one event and let it stand for all others, whether inside one organism, between two organisms, or among thousands of organisms. This event occurs with some variation in all chemical communication. It also occurs in communication involving touch, light, or any other material carrier. What's important is that this event and others are themselves a form of communication. Thus, the epic told by all the above examples involves innumerable subplots, nested one inside another, an exquisite set of precisely crafted Russian dolls, inconceivably complex.

The example we will examine involves a tissue—a group of tightly bonded cells—thickening in response to a chemical signal. We encountered this example before, as part of the conversation creating an eye. (The neural tube sends a chemical command across the minute, fluid-filled space between tissues; to the ectoderm, this command means "thicken!") The thickening of a tissue can occur in at least two ways. First, cells can divide and thus make more of their kind. Second, cells can swell, taking up water by opening tiny gates in the cell wall.[34] Both involve the same essential first step, the arrival of a signal at a cell's surface.

The cell surface is an exceedingly thin membrane insulating the cell from the outside world. It is studded with thousands of large molecules called receptors that consist of many smaller building blocks. Aside from receptors, billions of other molecules occur in and around the cell membrane. Some, like water molecules, are tiny, and others are hulking proteins with thousands of atoms. It is a crowded world, more densely packed with molecules than a subway at rush hour with people. Every instant, billions of molecules bounce into the cell membrane, many of them into receptor molecules. A tiny fraction of these molecules are the molecular messages telling a cell to thicken. When one of them bounces into its receptor, the receptor changes shape. This shape change is a key step linking one cell's message and another cell's response. But how does a receptor distinguish that chemical message from other molecules without meaning? Part of the answer is that these chemical words have a peculiar shape, a shape that is a negative—a bit like a plaster cast mold—of part of the receptor.

In some countries, people celebrate the New Year's arrival by pouring drops of searing, molten lead into cold water. Folklore has it that the future reveals itself in the congealed lead's bizarre shapes, no two of which are ever the same. Similarly, many molecules in and around a cell membrane have exceedingly complicated shapes, no two of which would match by chance alone. Thus, a molecular word arriving at one cell from another tissue would only match its own receptor's shape. Message and receptor are a molecular lock and key more sophisticated than any human-designed lock and key. (And shape is not even the whole story; other features such as electrical charge of message and receptor may also need to match for a response to occur.) The response, the receptor's changing shape, may couple the message directly to the cell's swelling. For instance, the receptor itself may be a channel with a valve that opens by changing shape. Before arrival of the right chemical message, this valve is sealed hermetically. On receiving the message the valve opens. Through the channel, small molecules like ions or water may enter the cell. And the cell swells.[35]

Matters are more complicated where tissues thicken through increasing the number of cells. In that case, receptors are only one of many links between message and response. Being large molecules, they protrude from the outer surface of the cell membrane, where they can respond to messages from other cells. And they protrude into the cell. Like the cell's exterior, the cell's interior is densely packed with molecules big and small, all bouncing into one another and into the cell membrane. After receiving a molecular message, the whole receptor shifts shape, including those of its parts that protrude into the cell. Some of the countless molecules inside the cell match the receptor's changed shape. And when bouncing into the receptor, these molecules themselves respond. For them, the changed receptor, whose shape they now recognize, is itself a message, meaning "change shape." And they change their shape. And yet other molecules recognize their changed shape and respond in kind. Hundreds, perhaps thousands of these communications are necessary before a cell divides to make more of itself. The basic principles, however, are the same: one molecule recognizes the shape of another and changes its shape.

Not all of the millions of receptors that pack the surface of a cell are identical. That is, the cell surface contains thousands of kinds of receptors, each kind with a different shape, and each recognizing one particular molecular message. And the number of molecules for any one kind of receptor may itself be in the thousands. One kind of receptor recognizes and matches one kind of message. Others do not match this message. They cannot recognize its meaning. Instead, these receptors recognize meaning in other molecules. They are responsible for receiving others among the thousands of messages any cell can receive and respond to. And a cell's responses are no less varied than the messages themselves. They range from the transmission of signals to other cells to the movement of the cell through the body to the cell's complete self-transformation and even to a cell's suicide. In other words, cells recognize a sophisticated molecular language involving shapes. This language is vastly different from the language you are presently reading, but it is at least as complex and powerful. It brings forth all of life.

I compared receptor and message to lock and key. This comparison, however, is as seductive as it is misleading. One generally cannot predict whether a receptor responds to an individual molecule, even if that molecule is the message intended for it. In other words, if you were able to follow an individual chemical word on its erratic path to a collision with the matching receptor, you would not be able to predict whether the receptor would respond. All you could predict is the average response of many identical receptors to the molecule. That is, most receptors would change shape, but you could not be sure what any one of them would do.[36] And a cell's response as a whole would be even less predictable. A message may be arriving. Receptors may be in place to receive it. Yet the cell's response may depend on previous messages and on its response to them.[37] How individual cells respond to a message may even be unpredictable in groups of identical cells, such as cells in one tissue. Usually, all you can predict is how most of the cells will respond. Thus, even in this microcosm of molecules, we find a principle familiar from other forms of communication, be they between human or nonhuman: the uncertainty of how the other will respond. The analogy of key and lock is thus flawed in that it insinuates a mechanical, machinelike perspective of molecular communication. The intrinsic uncertainty of communication, however, prevails even when individual molecules communicate.[38] Only the average response of many receptors or cells is predictable.

In sum, the link between communication among cells or organisms and their material transformation — into swimming fortresses, eyes, or parasitic plants — itself involves communication. In this form of communication, the messages are molecules released by a cell, tissue, or organisms. Their recipients are other molecules, or receptors. And each molecular message stands for something, means something to its receptor. At the very least, it means "change shape."

Dare we view molecules as communicating? You could choose to reject this perspective, a perfectly sound choice. It would make sense especially if you were interested only in conversations between organisms, what such conversations have in common, and what distinguishes them from other events in the world. That is, it would be a choice useful for a purpose. (But it is also no more than that, a choice of perspective. Like all choices of perspective, it does not necessarily reflect a truth.) The opposing choice calls the interaction between receptor and message a form of communication. As with any choice, we will lose something in adopting one perspective over others. For instance, we will lose the distinction between communication among the living and the nonliving. But, as with any choice, we will also gain. At the very least, we will gain a complementary yet equally all-encompassing perspective of the world around us.

The Nonliving in Communication

To view molecules as communicating we must allow the nonliving to participate in communication. Here are a few observations in favor of doing so.

Even in the common, narrow sense of communication, the nonliving plays an important role in both message and response. Take humans. A footprint, a fleeting pattern casually left in the sand, revealed the presence of another human to the stranded Robinson Crusoe. Important areas of science, such as archaeology, depend on similar signs. Archaeology's success in reconstructing the past depends on signs left by our deceased and decomposed ancestors. In skilled hands, these signs can tell a riveting story of life many millennia ago.

Nonhuman communication provides many other examples. Animals and plants read cold nights and short days as a sign of the approaching winter. They respond in ways as varied as hibernating and shedding leaves. Starlight is a sign to migratory animals, a beacon guiding them over thousands of kilometers. In these examples, the nonliving sends a message. But the nonliving can also respond. Take satellites, unmanned space stations, or planetary probes. Ground controllers can signal them to change their orbit, to collect rocks hundreds of million kilometers away, or to self-destruct. More pedestrian examples include TV remote controls and answering machines. And finally, none of these conversations need to be one-way. Machines can maintain a building's environment: ventilation, heating, cooling, humidity, lighting, and so forth. Although created by humans, they do so without human intervention, through a conversation between thermostats, motion detectors, hygrometers, and computers, to name a few. Communicating computers — sometimes many thousands of them — can jointly perform complex computations without any human intervention, and often much better than with such intervention. Familiar themes recur here, including miscommunication, unintended signals, eavesdropping, and parasitism.[39] In sum, the nonliving is centrally involved in many conversations, not only as one of the countless links between message and response but also as a partner in conversation.

Science as Conversation?

Science is a conversation between humans and the rest of the world; it is a form of communication where humans ask a question and the world responds. This conversation of science has shaped our world over hundreds of years like no other. For this reason, it plays an important role in this book.

Let's first consider the social sciences. These fields deal with people and include sociology, economics, and political science. Social scientists are both worse off and better off than biologists, chemists, and physicists. They are worse off because they cannot easily manipulate what they study — manipulating an economy or a society could easily wreak havoc with people's lives.[40] In other words, unlike physicists, who can smash elementary particles, social scientists must rely heavily on observation. But social scientists' disadvantages are compensated by an advantage. Economists, political scientists, and sociologists can usually observe more easily and directly than natural scientists, especially those who study subatomic particles or distant galaxies. They can simply ask questions of governments, companies, and — ultimately always — people. These questions then become part of a conversation, a conversation among many. Let's imagine a social scientist interested in the role of society — of agreement and disagreement among people about theories or facts — in acquiring scientific knowledge. He or she might poll people's opinion by asking the following question: "Agreement among people about reality does not influence reality and thus the course the world takes. Do you agree or disagree?" After collecting the answers, he or she might dig deeper and pose the next question: "Polls like this one have little influence on what people agree upon. Do you agree or disagree?"

It is not hard to see this as a conversation. But what about the natural sciences? Imagine this very simple example, a biological experiment. There is a south-facing kitchen window in my house. Its windowsill is crowded with houseplants. One day I observe that all their leaves line up in parallel with the window. This is my point of departure for a conversation.[41] I begin by asking a basic question: Why? Why do the leaves face the window? It occurs to me that they might be attracted by the window itself or by something outside it. So I turn one plant away from the window. This is my way of asking the world a question: Does the window attract the leaves? Some time later I come back and find that the leaves have turned toward the window. The answer was yes. And I might repeat this game a few times, perhaps with other plants, each time finding that the answer is yes. But now the next question rolls into my head. Are the leaves attracted by the window's glass itself or by something outside the window? It occurs to me that they might be attracted by my car, which is parked right outside the window. So I park the car elsewhere and turn one plant away from the window. Sure enough, its leaves turn back toward the window. The answer to this question was a resounding no. This conversation could continue indefinitely. It would lead us to the answer we all know, that leaves are attracted to sunlight. Similar conversations have been taking place for hundreds of years, involving many generations of scientists in physics, chemistry, and biology.

If one thing unifies scientists, it is that they pose questions to the world and receive answers in return. A good conversationalist's next question will depend on previously received answers. These answers in turn depend on the questions asked. And answers come in many shapes, a plant turning its leaves, people returning questionnaires, an embryo not developing properly, the contents of a test tube changing color, or an electronic blip, standing for a subatomic particle, appearing on a computer screen.[42]

Changing Conversation Partners

To be able to view science as conversation, we must accept two things. First, we must accept that the nonliving can play a role in communication.

Second, we must accept that communication can change those who communicate. To be sure, some kinds of communication may not change the participants much. Other kinds of communication—I have mentioned examples—create organisms or even change the course of the world. Whether minutely or profoundly, all communication mentioned so far changes its partners. This is perhaps most obvious for the last two examples, the conversation of shape-changing receptors and that of science. When I turn a plant away from the window, I change the world, however minutely. And the plant's response both changes how I look at the world and influences the question I ask next. Similarly, the arrival of a molecular message changes its receptor. It also alters the fate of the message. For one thing, the molecular message may now be stuck to the receptor. In addition, after the receptor has done its job of shifting shape, the cell often chops it—attached message and all—into tiny recyclable pieces.

Receptor and message stand for innumerable similar events that occur every instant inside the living. Many of these events are chemical reactions that change the reaction partners more drastically than just in shape. One molecule or atom, call it A, collides with another, B. If A matches B in shape, electric charge, or some other way, then the two react and are transformed into C. And what holds for chemical reactions inside the living is true for all other chemical reactions, whether in the atmosphere, in a garbage incinerator, on the bottom of the sea, in a fertilizer factory, or in the heart of a distant star. It is also true for the smallest microcosm we know anything about, the realm of such subatomic particles as electrons and positrons. When the right two particles collide, they annihilate and something new—a new particle, perhaps, or electromagnetic radiation—emerges.

There is nothing original or unusual about these observations. Less usual is the view—as in the case of receptor and message—that B has meaning to A, and vice versa. That this meaning is an instruction to undergo a certain transformation. And that the response is this very transformation.

The main problem with this choice of perspective is that we are used to thinking of such events in terms of cause and effect, as one event causing another, forcibly and predictably, like the gears and wheels that drive a clock. If you have trouble standing in the other place, a place where an immense conversation creates and sustains the world, this habit is probably at the root of it. You, me, and our ancestors have been completely immersed in a different worldview, where meaning exists only in us. Before we were sufficiently aware to make a choice, it had already been made clear to us that the world was an immensely complicated machinery but no more mindful than, say, a toaster. People committed to this perspective could concede that many events may look like conversations, that the partners in the conversation change, that the conversation's outcome depends on both partners, and so on. But they might call all these similarities "superficial." They say that in truth the world is a giant causal machine. That is certainly the easy perspective to choose. But we have other choices. And nothing — except well-worn habit — prevents us from taking the opposite perspective, that meaning and not matter rules the world.

Let me point to an important development in twentieth-century science that speaks to this issue. This development exposed fundamental flaws in a perspective that viewed the world as a clockwork of tiny submicroscopic gears and wheels. In fact, not only the clock but also its gears and wheels all but disappeared with the advent of quantum physics. Physicists found that the nature of the smallest things in the world, be they wave or particle, depends on the question we ask of them. It is the outcome of a physicist's and her machinery's conversation with these "particles." Perhaps a worldview rooted in meaning may be fundamentally flawed. But if so, a worldview rooted in mechanics is equally flawed. For if we look closely at its parts, they melt away into a haze of uncertainty.[43] Thus, if a mechanical worldview is to be favored, it is not because it is better founded than other perspectives.[44]

But when to choose one worldview over another? When that choice serves a particular purpose in a specific conversation. Do not dismiss mechanical worldviews; they are great at explaining how matter behaves in the world, and if our conversation were about building earthquake-proof houses, producing fertilizer, or designing thermostats, the mechanical worldview might be the most effective choice. But like an eye that cannot see itself, and like all views of the world, this perspective has fundamental limitations, blind spots. One is that it leaves no space for meaning.

These blind spots are where worldviews come undone.[45] To avoid them, we can only take a complementary perspective, a perspective that is strong where its counterpoint is weak. The perspective I advocate here is such a complement, the other side of the coin. To be sure, it has blind spots as well (blind spots are inevitable if you want to see). But as opposed to the mechanical perspective, it has been neglected for centuries. Perhaps the time has come to explore the world from this viewpoint in earnest, to see where it might take us while never forgetting that it, too, is merely a perspective.

It is tempting to deride a meaning-based worldview as anthropocentric, to distinguish it from worldviews seemingly less so. Conversations, communication, and signs, one might say, are exquisitely human. They are what connect us to the world. Molecules, atoms, and subatomic particles reveal their presence through signs, a flash of color in a test tube, an electronic trace in a detector, a blurry picture in a powerful microscope. Even our most immediate experience of the world depends utterly on signs. Take my experience of writing these lines. However immediate to me, the act requires millions of microscopic conversations, even if you count only those necessary for my fingertips and my brain to communicate. To build a worldview around meaning may thus appear to be the most anthropocentric endeavor of all. But watch out. Any other perspective, even a mechanical one where particles crash into one another in an endless, mindless serial accident of cause and effect is equally "anthropocentric." Why? Because it rests on metaphors derived from daily experience (which involves signs and conversation), such as billiard balls, gears and wheels, and water waves. And even concepts central to any perspective such as cause and effect originate in everyday events. As the philosopher David Hume pointed out in the eighteenth century, we always infer causes from effects through their association.[46] Effects stand for causes; they are signs for causes.[47]

Our everyday world is all we have. What lies behind also lies behind all appearances. It is a world from which even the meaning of "world"—let alone "matter" and "meaning"—have been removed. We can say nothing about it. Absolutely nothing. Thus, if a meaning-centered worldview is anthropocentric, so is any other worldview. Yet even acknowledging this leaves us with innumerable choices, choices of perspectives best for a particular purpose, choices that give us power to understand some aspect of the world and others that leave us powerless.

The perspective I advocate here is not simply anthropocentric for another reason. Even if no humans were around, signs, meaning, and conversation would still exist. To be sure, this perspective centers on something. But this something is certainly not human. I would thus rather call this perspective logocentric. It centers around meaning and mind — that which detects meaning in its most general form. In this perspective no thing, no matter how utterly different and separate, stands by itself. Everything stands for, refers to, and relates to something else, an infinite network of signs.[48] Signs are how the world speaks to itself, in an endless inner dialogue of creation.[49]

CHAPTER 2

The Other Side
of Self

Love one another but make not a bond of love:

Let it rather be a moving sea between the

shores of your souls.

Khalil Gibran

The time was August 1941. In the Auschwitz concentration camp, a prisoner had escaped from barrack 14. The reprisals for such escapes were draconian: for each escapee, ten men were condemned to death. The six hundred men of barrack 14 assembled. The assistant commander of the camp, SS Karl Fritsch, walked the rows of prisoners. Whomever he asked to step out of line would suffer death. One of the chosen men, Francis Gajowniczek, father of a family, wept and pled to be spared when an eleventh men stepped out of line. This man, the Franciscan priest Maximilian Kolbe, offered his life in place of Gajowniczek's. Kolbe was locked in a bunker with the other nine victims, left without food or water, to die a slow and gruesome death. The life of Gajowniczek was spared.[1]

Such striking examples of human self-sacrifice are rare. But more mundane acts of altruism are part of everybody's life. Parents, for instance, sometimes make astonishing sacrifices for their children. Why? If we asked them, they might answer simply that their children's well-being is their own happiness. Is their behavior selfless?[2] Many parents would say yes. Many biologists would say no.

And what about teachers? They might simply take pleasure at watching others learn and grow. Or they might cherish conversations with their students, where insightful questions sometimes help them see their field in a new light. In the end, teachers may be teaching for themselves as much as for their students.

Self-Serving Altruism?

Do people truly do things for others? Or do they ultimately always serve self? Such self-service may be exceedingly subtle, such as the pleasure of learning through teaching, of seeing somebody else happy, or of having sacrificed for a greater good. If scrutinized, this simple question about altruism leads to confusion, with no universally accepted answer. (Even sophisticated psychological experiments have not yielded one.)[3] But although this question matters especially to us, to our hopes and dreams of what the world should be like, the drama of self and other plays on a stage much grander than the human arena. For the distinction between self and other arose with life itself. It applies to any two organisms, whether animal or plant, weighing several tons or microscopically small, walking on land, gliding in the air, or swimming in water.

Does any organism, knowingly or unknowingly, act on behalf of others, disregarding self? Is there genuine altruism among the living? This chapter offers an overarching perspective on some astonishing behaviors, such as animal parents that slave their lives away for their brood, glue-grenade ants that blow themselves up for their colony, plant cells that give up life itself for a greater whole, and many others. This perspective harbors a paradoxical tension: the answer to the opening question is not clean-cut. Consider that self (you) and other (I) are like two sides of the same coin. Viewed from one perspective, they are completely separate. Viewed from another, they are completely inseparable. No one viewpoint is more valid than the other; hence, self and other share a paradoxical relationship. The question of altruism can be answered only by choosing a side in this paradox, however limited this choice may be, however inadequate and ultimately wrong.[4]

Let us first consider parents and how they behave toward their children and their kin. Why? Because human parents often feel that they behave altruistically toward their children and because biologists have studied parental behavior extensively.

Juvenile birds devour food ravenously, greedily, and incessantly. Their parents thus scramble ceaselessly to procure it, whereas they might actually prefer to improve the nest (or just go have a worm for themselves). Where such parental sacrifice is measurable, it is impressive. The large, ground-dwelling mallee fowl of Australia buries its eggs in large mounds made of soil, twigs, and leaves. The mound not only protects the eggs it harbors but, because it contains decaying compost, generates heat for the developing embryos. For more than eight months, the male mallee fowl will spend up to five hours a day monitoring the temperature of the mound, shifting more than 850 kilos of building material, the equivalent of forty tons for a human of average weight, to maintain the optimum temperature for the developing embryos.[5]

This behavior is not, of course, the ultimate self-sacrifice. Who has not heard stories of birds luring potential predators away from their nestlings, pretending that they are easy prey, and thus risking death? Animals that live in colonies have developed especially daring devices to protect their offspring and their colonies. California ground squirrels, for example, hurl sand at rattlesnakes that approach their underground nests. They and many other species also issue shrill alarm calls when assailants — whether airborne, terrestrial, or amphibious — approach. The warning call means "food" to the predator and thus often spells death to the caller.[6] Such self-sacrifice is taken to the extreme by glue-grenade ants: in defending their colony, they become suicide bombers, blowing themselves up by violently bursting a large gland and covering intruders with a disabling glue.[7]

These examples illustrate two important points about altruism. First, a behavior that does not cost anything is not altruistic. Cost can take many forms, some of them subtle: cost of energy, as for nesting birds; cost of time, as for a human parent giving up a career; or cost in the form of taking a risk, as for alarm callers and suicide bombers.

Second, organisms do not behave altruistically toward just anybody. Soldiers — whether ants or humans — are willing to die for colony or country but not for the enemy. And parents work hard to raise their young, but not just anybody's young — at least knowingly.[8] This very observation is key to the perspective that dominates biologists' thinking about altruism.

But before getting to the essence of this perspective, I should review how most biologists think about evolution, natural selection, and inheritance.

Species, Populations, Fitness, and Evolution

Organisms inevitably need to deal with—*interact* with—other organisms and with the rest of their world. Some interactions appear to benefit the organisms involved. Plants interact with sunlight, which gives them energy they need to grow. I interact with my lunch in a similar way. Parents interact with each other and their children to raise them. Each such interaction, looked at carefully, has a second, darker side. This darker side is sometimes glaringly obvious, such as in the case of the praying mantis. Here the female, copulating with the male, begins to devour her male partner, head first, while still mating. (The female literally makes the male lose its head, which is unnecessary for mating.) The female cooperates in copulation but at the same time treats her partner as food.

Not all interactions between organisms make such a striking comparison, but most have two features in common. First, two organisms usually have conflicting and converging interests—at the same time. Just think of the relationship between parents and their offspring—whether human or not. Time and energy devoted to offspring is time and energy lost to oneself. Second, viewed from the right angle, what is good for one is bad for another. Consider a seemingly innocuous interaction: plants extracting energy from sunlight, an inexhaustible resource. For what could that possibly be bad? For other plants, because plants compete fiercely for sunlight. In a tropical rain forest, the leaves of tall trees form a closed canopy, a huge, many-staked tent spreading from horizon to horizon. Its canvas filters most sunlight before the light can illuminate the ground, resulting in a dimly lit forest floor. And only plants that need little light survive there. All others die. More generally, all resources—light, water, nutrients, prey, soil, even time—are ultimately limited, and the benefit of self thus is the detriment of (some) other.

One type of interaction between organisms is special: sex, an interaction serving reproduction. Sex is closely linked with the notion of a *biological species*. By definition, two organisms that can reproduce with each other belong in the same species. But to reproduce, two organisms usually must get close to each other. If a species of flies occurs both on the mainland United States and on Hawaii, separated by an immense distance of open ocean, mainland and island flies will not usually get together, much less reproduce with each other. Do they belong to different species then? No. They belong to one species but different *populations,* groups of organisms living close enough to reproduce. How close is close enough? That depends both on the organism and on the world around it. Flowering plants do not travel, and even a wide river may cut a fly population in two. Birds, in contrast, may fly hundreds of kilometers, reaching beyond mountain ranges, large lakes, or deserts. Their populations may cover vast areas.

But organisms in one population not only reproduce with each other. Most of the time they compete, and even more fiercely with each other than with other organisms. Why? Because their ways of making a living, their lifestyles, are alike. Plants in the same population compete for sunlight, animals compete for food, microbes in the soil compete for nitrogen and carbon. And whenever a limited amount of an essential good is available—sunlight, food, minerals, water—competition ensues.

How does competition affect the living? To answer this question, let us examine reproduction. Reproduction is costly, and by that I mean not only the cost of raising the young. To produce seeds, for instance, plants need energy and building materials, including carbon dioxide from the air, minerals, and water. How much offspring a plant can produce depends on how much of these resources it can commandeer. Now suppose two plants of the same species live next to each other, one of them partly shading the other. The shaded plant has access to less sunlight for producing energy, and thus for producing seeds and offspring. What determines whether one plant shades another? Well, it may just be one plant's luck that it started to grow at that site a few years earlier. However, one of the two plants may be better at exploiting resources. It may be better at extracting minerals from the soil, like a mining company with more advanced mining techniques. With more building materials it grows faster. Even if the two plants germinated on the same day of the same month of the same year, the plant with the better mining techniques will grow faster. Eventually, it will shade the other plant and reduce its access to sunlight.

Whenever the abilities of two organisms to survive or reproduce differ, biologists say that these organisms have different *fitness.* In the above example, two plants differ in their ability to *reproduce,* but their ability to *survive* is just as important. One of our two plants, for instance, might succumb to droughts more easily, or it might perish in salty soil.

Any feature of an organism may affect its ability to survive and reproduce in many ways. Animals may differ in how they fend off parasites; plants in the molecules they produce to poison their enemies; poison arrow frogs in their coloration deterring predators; plants in how they show off their flowers to insects; carnivorous plants in how they lure their prey; spiders in how they choose a site for their webs; large predators in how skillfully they lay an ambush; mosquitoes in how effectively they suck blood and get away with it; wind-pollinated plants in the ability of their pollen to disperse. The possibilities are endless. In fact if you think of any — and I mean any — feature of an organism, you could think of a way that this feature might aid in surviving or reproducing. Fitness is thus a powerful concept.[9] Nonetheless, like any concept and human creation, it sometimes fails. For example, it fails in explaining why any parent would help its offspring. Only a key modification permits that.

Another important concept is *heritability.* If two rabbits that can sprint faster than other rabbits have offspring, their offspring are often fast runners, too. If they are rather slow runners, their offspring may also be slow. There is, however, no way to be certain of the offspring's features. Fast-running rabbits may occasionally have slow offspring, and slow-running rabbits may produce world-class sprinters. But on average, fast-running rabbits produce fast-running offspring. Heritability is about this *tendency:* if offspring tend to be similar to their parents with respect to a particular feature, we say that the feature is *heritable,* or *inherited* from parents to offspring.[10]

Here, then, are the main ingredients of *evolution by natural selection.* Evolution requires a *population* of organisms. There have to be *differences* among these organisms *in their ability to survive and reproduce.* Whenever these differences are *inherited* from generation to generation, evolution can take place.

What happens when all these ingredients are in place, in, for example, a plant population? Some of the plants might extract soil minerals very efficiently, whereas others do not. The efficient plants grow faster. The fast-growing, efficiently mining plants often shade the slow-growing ones. Therefore the slow-growing plants produce fewer flowers and less pollen.

Not only that, they also produce fewer seeds. Overall, they produce fewer offspring. Over many generations, the seeds and seedlings would come increasingly from fast-growing plants. And because these seedlings also grow faster, and themselves produce more seedlings, fast-growing plants will eventually dominate the scene. This is the essence of evolution by natural selection in this population. It is the process whereby fast-growing plants imperceptibly replace slow growers over many generations.

Genes

A final bit of background material has to do with the causes of inheritance. What makes inheritance possible? Genes or genetic material passed on from parents to offspring would seem to be the obvious answer. If so, why have I not mentioned genes so far? Because nothing I have said so far requires genes. Inheritance does not require genes. Even life itself can exist without genes. The earliest life on earth probably did not have genes. Yet it had inheritance, or neither of us would be here. And there is such a thing as cultural evolution. It does not require genes either, but it does require inheritance.[11] However, inheritance without genes is the exception among organisms alive today. And as long as one keeps these caveats in mind, genes are important for inheritance.

Let us first look at the what and where of genetic material. Organisms consist of cells, the smallest units of life. Some organisms consist of only one cell, others, like you and me, of billions. Each cell contains an identical copy of the organism's genetic material. This material consists of many thousands of genes made out of DNA. If you like to think of a cell as a machine (some do — it is another metaphor as powerful as it is dangerous), then each gene contains the instructions to make one part of the machine. The most important of these parts are proteins. (We already briefly encountered one kind of protein, receptors.)

As I argued earlier, any matter can carry meaning, whether or not it comes in the peculiar patterns of ink on paper we call written words. How does DNA convey its meaning? DNA is a long, stringlike molecule that consists of four different building blocks called nucleotides. These building blocks are abbreviated by four letters: A (which stands for the chemical compound adenine), C (for cytosine), G (for guanine), and T (for thymine). The information DNA conveys lies in the sequence of the nucleotides A, C, G, and T. Of course, each of the four nucleotides itself has a shape, and thus DNA, to be precise, speaks a language of shapes, similar to the proteins we encountered when cells converse.[12] But because we are most familiar with languages that convey meaning in letters, we find it more convenient to think of DNA as a string of letters.

At first sight, a language of only four letters seems awfully simple. How many genes could one possibly make up in this language? A typical gene may consist of a thousand nucleotide letters or more. Some physicists believe that they can estimate the number of atoms in this universe. This number is very large, greater than a one with eighty zeroes behind it. It is, however, tiny compared to the number of possible genes of length one thousand. The language of DNA can produce more genes than you or I could imagine.

Because DNA and genes are key to making an organism's parts, they influence an organism's appearance and actions. Unfortunately, we do not know exactly how genes influence the more complicated features of organisms—the ability of their pollen to disperse, the brilliance of their plumage, the ampleness of their birth canals, or their generosity toward others—except that many genes contribute to these features. But we do know how individual genes are inherited, how they are passed from generation to generation. The cells of many organisms—ours included—do not contain one but two copies of each of the thousands of our genes. Biologists call these two copies of a gene *alleles*. However, some cells in our bodies carry only one copy, one allele, of each gene. These are sperm and egg cells, produced by male testicles and female ovaries. If a male and a female have a child, one sperm and one egg cell fuse at conception. The resulting cell, the fertilized egg, has two copies of each gene. This cell divides countless times, a process whose result is a newborn child with billions of cells. During each such cell division, a cell passes on an exact copy of all its DNA. As a result, each of the child's cells also has two copies of each gene, one from the father and one from the mother.

If we inherit all of our genes from our parents, why can we not predict a child's features and peculiarities from those of its parents? First, most of our cells—with the exception of sperm and egg cells—contain two alleles of each gene. Which of these two copies will make it into a sperm or into an egg cell? Either one can, but which one does is a matter of pure chance. In our body, one of the two alleles is chosen at random and placed into an egg or sperm cell. And the two copies, having been inherited from the organism's parents, might be different. Thus, every sperm or egg cell is different. How many different sperm cells could you produce by picking one of two alleles at random, for each of thousands of genes? The number is even greater than the number of thousand-letter genes, and thus much greater than the number of atoms in the universe.

Another reason for varied inheritance is that cells make copying errors when dividing. That is, when a cell copies a gene, it sometimes makes mistakes. The cell's job is to make an exact copy of a string comprising, say a thousand As, Cs, Gs, and Ts. But sometimes the cell might replace an A with a T. Or with a G, or with a C. It might skip a few letters, rendering the copy shorter than the original. It might add a few, making the copy longer. Mistakes like these occur only rarely. However, organisms like you and me have thousands of genes, each of which is copied every time a cell divides, and thus many times between fertilization and birth. Even rare errors can thus accumulate and render the genetic material of child and parents different.

Here is a final reason for differences among offspring, more important than imperfect copying and a random lottery of alleles: not only an organism's genes but also the world around it influences its features. And the world the parents experienced is different from the world their offspring experiences. There is no better illustration of this principle than identical twins, who have identical genes. Their biographies are usually different, not only in the careers they choose and in the diseases that befall them, but in the many fortunes and misfortunes that shape their life until they die, usually from different causes.[13]

The simple mechanism of inheritance involved in sexual reproduction, stuffing one of two gene copies into a cell and sending the cell on its way, has an important consequence. It allows us to estimate how closely related two organisms are *genetically*. For example, assume that you and I are siblings. And let's look at one of the copies of a gene, any gene, in your body. You got this copy either from our father (from the sperm) or from our mother (from the egg cell). Assume first that you got it from our mother. Now let's look at one of the two copies of this same gene in *my* body. (It does not matter which copy.) What are the odds that I, your sibling, have also inherited this copy from our mother? If you recall that our mother's body simply places one of the two copies into each egg cell at random, the odds are 50 percent (because chances are 50 percent that I have gotten *any* copy of any gene from our mother). But what if *you* got the copy of this gene from our *father?* Well, then, what are the chances that I also got that copy from our father? They are 50 percent, for the same reason. Regardless of your perspective, the chances that a copy of a gene, any gene, inside my and your body stem from the same copy in our parents is 50 percent. In this sense, siblings are 50 percent genetically related. A similar argument works for other relatives, including parents and offspring (they are also 50 percent related), grandparents and grand-children (25 percent), cousins (25 percent), and so forth.

This procedure of calculating genetic relatedness is far from perfect. It works strictly only one gene at a time and calculates only average related-ness. For example, even two siblings that are not identical twins might have gotten all the same gene copies from their parents. This is possible although extremely unlikely, more unlikely than winning the lottery every single week of one's life. In this case, the siblings would be 100 percent related, and the procedure would have erred by predicting that they are only 50 percent related. But despite its shortfalls, this procedure is often extremely useful, for example in predicting the odds that somebody will carry a genetic disease.

Back to Self and Other

Equipped with all this information, let us return to self and other. First, altruism is behavior benefiting another but costing you. If costly behavior is bad, then altruism is bad for self and good for the other. Second, non-humans and humans alike do things for some other organisms, especially for their offspring, but not for all other organisms. How do they choose their beneficiaries?

To begin with, this question would probably not even occur to most people, who would consider the existence of true altruism self-evident. But to biologists, altruism is by no means self-evident. For more than a century, many — scientists and nonscientists alike — have been taking a particular perspective on the living. Let's call it the tooth-and-claw per-spective. You are probably familiar with its catchphrases: "everybody is in it for themselves," or "the bloody struggle for survival." This perspective is different from the mere claim that natural selection has shaped the living.

For this perspective expects, in addition, certain behaviors that are competitive, ruthless, and most of all selfish. From this viewpoint, it is indeed difficult to explain why anybody would do anything for anybody else. The question why there is altruism thus arises rather naturally from this perspective's blind spot.[14]

In the early 1960s, the biologist Bill Hamilton came up with an explanation of parental altruism and many other kinds of altruism.[15] This explanation consists of two parts: that organisms help each other when they are related — the explanation's core — and that organisms help others for only one purpose: to help themselves. The second part is a tacit interpretation of the explanation's core.[16] In this interpretation, altruism is selfishness in disguise. The core and this interpretation of it are usually not seen as distinct. But they are. And I will go to some effort to show that they are and that the tooth-and-claw perspective sees only one side of the coin: it does not see the other side of self.

Selfish Parents

To get to this point, I will now delve into some details of how many biologists think about altruism. To begin with, their answer about the roots of altruism rests on two premises. First, genes can influence an organism's behavior. The argument for this premise goes as follows. Different gene copies, different alleles, differ slightly. As a result, the protein parts made from the information in them differ as well. And these different protein parts can have different effects on behavior. The behavior in question here is selfish and altruistic behavior. In other words, some alleles of a gene — the selfish alleles — cause selfish behavior, whereas others — the selfless alleles — cause selfless behavior.[17] For any behavior, some individuals in a population may harbor selfish alleles while others harbor selfless ones. The second premise is to view evolution from the gene's and not the organism's vantage point. Thus, rather than considering how organisms change over many generations, the focus is on selfish or altruistic alleles in a population and on which of them persist over the generations.

To illustrate this, let's revisit parental altruism, say in a population of birds. Some parents in the population may harbor altruistic alleles. These parents provide for their chicks as best as they can. Others may harbor alleles that dispose them to eat the chicks' food themselves when they feel like it or to look the other way when a predator approaches the nest. What is the fate of these different alleles in the course of many generations? The extreme situation where a parent sacrifices its life and thus saves its offspring illustrates their different fates most clearly. An allele that makes a parent sacrifice itself in the right moment will survive in the offspring. On the other hand, an allele that makes the parent look the other way will not make it into the next generation. Thus, alleles that make the parent sacrifice itself at the right moment will have a better chance of being passed from generation to generation. Selfish behavior will benefit the parent in the short run, whereas altruistic behavior will not. But altruistic behavior will also help the allele responsible for this behavior, whereas selfish behavior will not, because it brings about the offspring's death and, with it, the allele's extinction. Altruistic behavior from the organism's view becomes selfish behavior from the allele's point of view. Altruistic alleles are selfish alleles.

This example also illustrates that we must be careful to distinguish properly between self and other. Which self and which other is this about, parent and offspring? No, a different relationship has now become important, that between a gene (self) and the organism (other) whose behavior it influences. Viewed from this relationship, selfish alleles cause a parent's altruistic behavior toward its offspring.

To drive this line of thought further, which behavior would the opposite kind of allele, an altruistic allele, cause? An altruistic allele helps its host organism at a cost to itself. In other words, it causes the host to act selfishly. In that case, the host's offspring will perish, and the allele will pay the ultimate price: death. Thus, just as selfish genes cause altruistic behavior, so do altruistic genes cause selfish behavior.

This perspective on altruism rests on parents and offspring being genetically related. It explains why parents take care of their offspring but not of unrelated individuals in the population. Consider an individual that sacrifices its life for an unrelated other: any allele instrumental to this self-sacrifice would be doomed because it would perish with its carrier. It would not live on in the survivor, because this survivor, being unrelated, does not carry it. And conversely, an individual's selfish behavior toward others might go unpunished simply because these others — although they might perish as a result — do not harbor the allele responsible for this behavior.

But parents and offspring are not all there is to family. There are cousins, grandchildren, siblings, and many other relatives. Does the same reasoning work for them? Do they help one another because they are genetically related? They do. I have mentioned that one can estimate degrees of genetic relatedness among all of these relatives. The more closely related two relatives are, the more readily they make sacrifices for each other.[18] Others, especially the biologist Richard Dawkins, have written forcefully on this, so I need not dwell on individual examples.[19] Suffice it to say that this perspective on altruism becomes extremely powerful when applied to extended families. It can explain obviously altruistic behavior, such as why in colonies of bees, ants, and termites, individuals readily sacrifice their lives for the colony by hurling themselves against attackers many times their size or by blowing themselves up. It can explain why thousands of individuals in such colonies forego reproduction and why only one or a few queens lay eggs. It can explain obviously selfish behavior, such as why parasitic birds such as cuckoos and cowbirds lay their eggs in the nests of other birds and why male langurs and lions kill other males' offspring. It can even explain phenomena that, on the surface, have nothing to do with altruism or selfishness, such as why organisms as different as Seychelles warblers and paper wasps can change their sex ratio, the proportion of female to male offspring they produce.

Because this perspective, in which genes selfishly promote altruism, is powerful, we tend to forget that it might be just a perspective and get caught up in it. Here are two of its limitations, two leaks in the roof of the house science built. Their dripping reminds us that the world is less comfortable than our theories would have it be. First, individual genes usually do not make it intact through many generations. If you follow an allele through the generations, it will change through the copying errors cells make. How can one speak of a gene making an organism act on its own behalf if the gene itself changes over time? When are two copies similar enough to be called identical, to still be "self"? We don't know. An even more serious problem is that virtually no behavior is influenced by just one gene. Many, perhaps thousands of genes, influence behavior. And biologists find it excruciatingly difficult to pin down even the dozen or so most influential genes for most behaviors.[20] Although individual genes may pass intact from one generation to the next, this can hardly be said for multiple genes. Just recall that organisms scramble their genetic material every generation by placing one of two alleles into their sperm and egg cells. The selfish gene perspective loses its power if many genes influence behavior, because these genes are scrambled every generation.

Immortality through Death

Before leaving genetic relatedness, I want to relate a stunning last example of its power. The example is about events that first occurred some two thousand million years ago and that had the most profound impact on all of life. Not you, I, or any other human — or our conversations — would be around without them.[21]

Early life on earth did not involve many-celled organisms like us; rather, it consisted of single-celled organisms akin to bacteria.[22] The life story of such organisms is rapidly told: they grow and divide.[23] More precisely, a single-celled organism divides into two daughter cells after having grown to a certain size. The daughter cells are initially much smaller than their mother cell. But they keep growing until they reach a certain size. Then they divide, and the cycle repeats itself.

In a population of single-celled organisms, some may grow a little faster than others. They might extract nutrients from their surroundings more efficiently, like plants with better mining technique, or they might fare better in a hostile environment and be more resilient to cold, heat, or drought. Such features are often heritable. That is, if an organism grows rapidly, its two daughter cells will also grow rapidly. Perhaps not exactly as rapidly, perhaps a little faster or more slowly. But in general, fast growers will produce more fast growers.

How fast can fast growers grow? That depends on both the organism and its surroundings, but very fast growers might spend as little as a few minutes between cell divisions. If so, a single hour can see several generations of them. (Just compare that with the twenty-plus years of a human generation.) How does a population consisting of fast and slow growers change over time, over many generations? Simple. Fast growers take over the population precisely because they grow faster. They do not even have to grow much faster. Let's say that a slow grower needs one hour of growth between cell divisions and a fast grower fifty-nine and a half minutes. That is about 1 percent faster. In a population consisting of 50 percent fast and 50 percent slow growers, 99 percent of the population would consist of fast growers within less than a month. To persist in such a population of single-celled organisms, it is not only necessary to survive and to reproduce. It is also important to do it quickly.

For over one and a half billion years after life originated, this game of growing and dividing was the only game in town. But then, about two billion years ago, multicellular organisms appeared and changed the rules. They differ in many respects from single-celled organisms, but one difference is critical. When a multicellular organism — like you or me — dies, most of its cells die with it. The only cells that do not perish, that are potentially immortal, are sperm and egg cells. A few of them might participate in the making of a new generation and become immortalized in a chain of offspring. All other cells die.

Now put yourself in the position of a single-celled organism that will persist only if it divides rapidly. Consider that its perspective on life is: divide fast. What would it have to give up to become a part of a many-celled organism like you and me? First, it would have to give up dividing.[24] But even worse, it would have to give up life itself. What in the world could make it do that? Relatedness again provides the answer.[25]

A family of algae alive today, the Volvocaceae, dramatically illustrates the transition from single-celled to many-celled organisms.[26] These algae consist of several species of microscopically small organisms, much smaller than the tangles of seaweed most of us know as algae. These algae thrive all over the world in nutrient-rich fresh waters, from tiny puddles to vast lakes. Their cells are green, and each cell's surface anchors two whip-like protein spirals that turn repeatedly and very rapidly. These spirals are called flagella, and they serve as propellers, allowing the cell to swim.

Like other green plants, these algae procure their building materials from sunlight and from simple building blocks like carbon dioxide and minerals. They mine, however, not the soil but the waters through which their flagella propel them. Part of their survival strategy consists of scouring their world for new resources they need to grow. Think of them as mobile mines.

Each species in the Volvocaceae contains strikingly different organisms. Some species consist of single-celled organisms that reproduce through cell division. Other species consist of four cells held together by a jellylike glue known as mucilago. How do these colonies of four reproduce? Each cell in a small colony — or is it an organism? — grows very large. Then it divides rapidly not once but twice, thus producing four daughter cells. This division results in four groups of four cells, also held together by mucilago. The four groups then separate into four new colonies.

Other species in this group consist of sixteen cells, yet others of thirty-two cells. They reproduce similarly: each cell grows very large and then divides rapidly into sixteen or thirty-two daughter cells.

Division and colony formation changes radically in other species of this family, in which each organism contains up to two thousand cells. The cells are still the same, but the colony is now a behemoth with a volume more than a million times larger than each cell. It resembles a hollow balloon whose surface is covered with cells. Under the microscope, these giant balloons roll leisurely — almost majestically — across the field of vision. Their cells are presumably on the surface, because there the cells can propel the giant colony through the water with their flagella.

These giant colonies can no longer reproduce by simple division because they have too many cells. Instead, each balloon now consists of two cell types, cells that reproduce and cells that don't. The surface cells propelling the organism do not reproduce. Cells that do reproduce are set aside inside the balloon. They form little balloons inside the big balloon. There may be dozens of such little balloons for each big balloon. The reproducing cells grow and divide, until the little balloon they make up gets bigger and bigger and consists of more and more cells. At some point during the life of this organism, the big "mother" balloon ruptures, releasing all the little balloons. *The big balloon dies.* The little balloons grow into big balloons with little balloons inside, and the cycle repeats. The birth of the little balloons means death for the big balloon. Birth has come to entail death.

In sum, in the larger species of the Volvocaceae, all cells that are not part of a reproductive balloon — thousands of surface cells — will die. They have given up life.

A single-celled algae gains immensely by becoming part of a bigger whole, which becomes clear from the lifestyle of these colonies. You can think of these algae as tiny submarines. In each twenty-four-hour period, they dive or rise to various depths depending on their needs. During the day, it is best for them to be close to the water surface, making maximum use of available sunlight. But most nutrients are not near the surface. They occur in deeper waters. So at night, colonies move to deeper waters to mine nutrients there. In a deep lake, a colony may commute more than eighty meters daily between the surface and the nutrient-rich depth.[27] Large colonies with their thousands of propeller engines have a great advantage in this commute. They are fantastic swimmers. Some may swim as fast as five meters per hour. If that speed does not impress you, remember they are microscopically small, and consider that their swimming speed would translate into more than fifty kilometers an hour if they were as large as, say, a rainbow trout.[28] In addition, larger colonies can store more of the nutrients they have found, partly because they have more space for them. And recall that reproduction requires resources. Large colonies thus can produce more offspring than colonies with less storage space. And finally, nutrient-rich lakes are not only full of algae but also full of dangers. They harbor many predators, minute animals that filter enormous amounts of water to get at their food, which includes tiny algae. So numerous are these predators sometimes that they could filter all of a lake's water within a few days. But they cannot swallow larger algae. Being part of a large submarine also protects against predators.

Large colonies thus clearly have many advantages over small ones. But this observation does not fully answer a key question: Do these advantages make up for the loss of life, the sacrifices that thousands of individual cells make during reproduction? What does an individual cell stand to gain if it loses life?

Relatedness provides the answer. The cells in a colony are — except for a few errors in copying DNA — identical. They are family much more so than any human family.[29] They are almost 100 percent genetically related, as opposed to, say, the measly 50 percent of human parents and their children. Even if one or a thousand individual cells die, nearly identical copies of them live on. And by becoming part of a more effective submarine, chances increase many times that some of them will make it. Self (the cell) and other (the organism) are two sides of the same coin, utterly different but linked through their common fate, a link forged again by their genetic relatedness.

Relatedness and Egotism

This example vividly illustrates the enormous power of genetic relatedness. Without it, multicellular organisms would not exist. In genetic relatedness, shared genes are the glue that holds fates together. Most self-sacrifice discussed thus far — from that of parental birds to that of rolling green algae — is derived from genetic relatedness. Every birth of an organism creates new genetic relatedness.

In genetic relatedness, even the sacrifice of life itself can be construed as a selfish act because it leads to propagation of a cell's genes. But viewing altruists as helping themselves is a narrow perspective on the question why there is altruism. A broader perspective is that altruists help others related to them, be they related genetically or by virtue of their fates. It is ultimately relatedness in fate that influences how self behaves toward other. For even without genetic relatedness, the fates of self and other may be intertwined, as we shall see. Whatever self does unto other will influence its own fate. Although self and other may live and die by themselves as individuals, their fates may be inextricably linked, twisted around each other so that putting the ax to one fells the other.[30] And sometimes, their fates become so closely linked that their boundaries become blurred: they become one.

Standing in this place — where the focus is on the relationship between self and other, and not on either one of them — has obvious benefits: it can deflate acrimonious debates over altruism. It can also take us much further than the selfish gene perspective in understanding altruism. But as usual, one has to give up something to change perspectives. It is that the distinction between self and other is more important than their relatedness; that one side of the coin is more important than the coin itself.

Relatedness in Fate

Most organisms are not genetically related, yet their fates are linked just as inevitably. Insects and flowers — a well-worn example — serve to make this point.[31] There is no question of genetic relatedness here: not only do insects and flowers belong to different species, but one is an animal, the other part of a plant. Animals and plants are as different as organisms can be. Yet every child knows about bees, hummingbirds, and butterflies. It knows that flowers give them food and that these insects allow the plants to reproduce. One partner depends completely on the other.

The lives of some insects and plants are conjoined much more closely than we appreciate. Fig trees and their pollinators, fig wasps, provide an example of how closely linked their fates can become. The flowers of sycamore fig trees — like those of other fig trees — are sheltered inside hard goblets of plant tissue that will eventually ripen into figs. These goblets have a tiny opening at one end. Female fig wasps try to squeeze through this opening. But the opening is so narrow that not all wasps succeed. Even a successful wasp loses her wings and sometimes other body parts, such as her antennae. This doesn't matter, because she will no longer need them. Once inside the fig-to-be, a singular activity consumes the female: she drills tiny holes into the plant tissue and lays eggs in these holes. While doing so, she pollinates the flowers and then dies. The first offspring to hatch from her eggs are male fig wasps. They immediately scour their prison for unhatched females that might be below the prison's surface. On finding an unhatched female, a male drills a tiny tunnel through the plant tissue and copulates with her through that tunnel. After copulating, the male gnaws a larger hole through the outside wall of the goblet and then dies. (His is not a glamorous life.) After the young females have hatched, they climb out to freedom through this hole and in doing so take some pollen with them.

The lives of figs and their wasps depend completely on one another. Not only do fig trees need their wasps to reproduce, but the figs will not ripen (nor will female wasps hatch) until a male fig wasp has chewed a hole through the immature fig. Nothing about these organisms' life cycle would make more sense by insisting on separating the benefit of self (fig tree) and other (fig wasp). And these partners are in no way genetically related, yet their fates could not be more tightly linked.

Equally tight relationships exist for other flower-insect pairs. Eliminate other, and self is doomed. But insects and flowers are only the most vivid examples of related fates without related genes.[32] Others are everywhere around us. They are so numerous that it is impossible to study an organism without also unearthing other, genetically unrelated organisms it utterly depends on. Some such fate-mates live next to one another, such as the aphids that excrete honeydew for ants and receive protection in turn. Other fate-mates live inside one another, like the microscopic bacteria that help plants like clover and alfalfa procure nitrogen from air.

Large and small fate-mates are often linked by chains of partnerships. Such a chain exists in wood-eating termites. To help them digest wood, termites rely on small single-celled microorganisms living in their guts. These organisms break wood into smaller building blocks that a termite's gut can digest. The termite's gut in turn supplies its inhabitants with food. If you examined one of these microorganism's surface, you would find it covered with small whiplike flagella. Similar to the algal flagella I mentioned previously, these propellers allow the microorganisms to travel around within the gut's goo. If you took a closer look, however, you would find something astounding. These flagella themselves are single-celled whip-shaped organisms that turn incessantly on the surface of their hosts, propelling them forward and receiving food in return.

Should we think of such partnerships as altruistic or selfish? An insect transports pollen, which benefits the plant but also costs energy. By definition, the insect behaves altruistically. A plant gives nectar to the bee, nectar whose production also costs energy. The plant thus also behaves altruistically.[33] Both insect and plant, however, also receive a benefit. So is their behavior selfish? Is their altruism selfishness in disguise? (Conversely, could you view selfishness as altruism in disguise?) Consider that it may neither be necessary nor helpful to make this distinction, for the fate of one is inseparable from the fate of the other. Were you to harm self you would imperil the other.

But, you might say, although life has entangled the fates of *some* genetically unrelated organisms, for many others the benefits of self and other are distinct and even opposites.[34] For one thing, organisms harm and kill each other to make a living. But even such organisms — each other's worst enemies — provide a valuable lesson about self and other and their conjoined fates.[35]

We have already encountered parasites, organisms that live on or inside others and exploit them for food and as breeding grounds. The benefit seems to be the parasite's and the cost the host's. Entirely.

But beware. For many parasites are not what they seem. Consider the tiny microbes *Trichinella spiralis* and *Trypanosoma lewisi,* causes of severe human diseases such as trichinosis. They also infect mice, but they affect them in curious ways: some infected mice grow fatter and live longer than uninfected mice. Apparently these mice — however heavily exploited — benefit from substances such as vitamin B1 produced by the parasites.

The protozoan microbe that causes malaria is another case in point. In some regions of the world where this disease is rampant, the infected suffer less from high blood pressure, a major risk factor for heart disease. And perhaps the most striking example is the tiny isopod *Cymothoa exigua* (a relative of crabs and lobsters), which enters the mouth of its host fish, the rose snapper. There it feeds on the fish's tongue. One of nature's cruel abominations, you might think. But long after the tongue is gone, this parasite remains hooked to the victim's mouth and serves as a replacement tongue, helping the fish capture prey.

Not all parasites necessarily service their hosts. Many probably do not. But even where exploitation is most ruthless, self often serves other. Like the praying mantis that devours its male while mating with it, self and other ferociously assail yet simultaneously serve each other. Strong ties conjoin self and other even in parasites — however frequent or rare — that only exploit their host. Imagine, for example, the perfect parasite: a perfect parasitoid wasp laying its eggs into caterpillars of one insect species. So perfect is this wasp that one year it tracks down every caterpillar of its host species and lays its egg inside each of them. Although it may thus have obliterated a population of butterflies, this wasp has produced enormous quantities of offspring. It has served self perfectly. Or has it? For where is its offspring going to lay its eggs the next year? Nowhere. All its offspring will die because the host is gone. By obliterating other, this wasp has obliterated self.[36]

Do such mutually devastating interactions occur in nature? They would be hard to detect, because neither self nor other would be around long enough to tell their story. One observation, however, speaks to this question: how parasites can change the host organisms they exploit.

The worldwide AIDS epidemic is a case in point. It was triggered by a virus that jumped from apes to humans somewhere in sub-Saharan Africa in the early twentieth century.[37] In apes, the virus is perfectly harmless, whereas in humans it wreaks havoc. The medical literature is full of similar examples where a parasite, on changing hosts, devastates the new host population, killing many or all individuals. If some hosts survive, host and parasite continue to interact over many generations. Over time, the parasite often causes less and less damage. Some of this damage may be inevitable. It may reflect a parasite's minimal need for resources, or it may be a by-product of how the parasite moves about. For example, diarrhea is unpleasant and often dangerous for the host, but it is an important (though unglamorous) means for parasites to move from one intestinal pasture to another.

Sometimes such an increased interlocking of parasite and host fates can be replayed in the laboratory—with bacteria.[38] Bacteria, common parasites of humans, have their own parasites. These are bacteriophages, viruses that infect bacteria, reproduce inside them, and then move on to infect other bacteria—either a bacterium's offspring or bacteria they have never encountered before. Jim Bull from the University of Texas has propagated bacteriophages for many generations within the same bacterial population. He has observed that, over generations, the parasites damaged their host less and less, allowing the bacteria to grow faster and faster even in their presence. Over many generations, parasite and host thus come to share a common fate. The less self damages other in making a living, the better off both of them are. Their fates become conjoined.[39]

Becoming One

How closely linked can the fates of two organisms become? Genetic relatedness is the closest kind of relatedness we encountered thus far. Hosts and parasites generally are not genetically related. They can, however, become genetically related. Not only that, they can fuse to become one.

Some cells live inside other cells, including human cells. Examples include Rickettsiae or Chlamydiae, the causes of typhus and trachomas (a kind of eye infection leading to blindness). These parasites divide inside their human host cells and rob them of energy or nutrients. Rickettsiae and Chlamydiae can no longer live independently of their host. If a parasite is utterly dependent on one host, killing the host would amount to suicide. Such parasites may thus be more closely related to their host in fate than others, say parasitic wasps and their caterpillars. (The wasps, after all, spend only a fraction of their life inside the host.)

Two cells can become even more closely related in fate than that. The most vivid example regards two cells whose relationship began at least one and a half billion years ago, probably before multicellular organisms arose.[40] This relationship may originally have been parasitic, like that of Rickettsiae or Chlamydiae. Around that time, a small cell began to live inside a big cell and either took advantage of the big cell—perhaps to obtain its nutrients—or was taken advantage of. Over time, the two cells became dependent on each other. Whenever one cell divided, the other needed to divide with it, such that killing the other would have amounted to killing self. The interaction of these cells continues to this day, but it is no longer parasitic. It has become key to life as we know it, because each of our cells harbors tiny structures that descend from the smaller cell. These structures are called mitochondria. After all this time, mitochondria still display hallmarks of being cells. They are surrounded by a cell membrane, possess their own DNA, and produce some of their own protein parts from this DNA.

How are our fates and the fates of our mitochondria linked? First, mitochondria produce most of the energy that our cells need. They have become our power plants and thus an essential part of our cells. Damaged mitochondria cause severe diseases or death. Second, mitochondria are always passed on through egg cells into the next generation. They cannot leave their host cells anymore. They obtain many of the building blocks they need to grow and divide from the surrounding cell. And third, many previously mitochondrial genes have become part of the surrounding cell's DNA. These genes, necessary to make part of the mitochondrion, are not part of the mitochondrion anymore. When the cell makes a mitochondrion's part from such a gene, the mitochondrion must then import the part from the cell. Through these three relationships, our cells and their mitochondria are as tightly linked as can be. They are inherited together, are necessary for each other's survival, and have even exchanged genes. They truly have become one.

The Price of Relatedness

These last few examples show that many genetically unrelated organisms are related in fate, even organisms that harm each other. The closer that relationship, the less separable self and other become. To insist on distinguishing altruism and selfishness becomes a hollow exercise when self and other have become one.[41] But especially with this example in mind, why does cooperation not arise more often from parasitism? There are more parasites than free-living species, perhaps many times more. Why is parasitism so abundant?

The concept of a parasite exists only by a collective human choice, an agreement on a definition: parasites take resources from other organisms and live close to, on, or even inside their host. But from a broader perspective, most life is "parasitic." Carnivores are parasites of their prey. Herbivores inflict great damage on their food plants. Even the many soil-dwelling organisms — animal, fungi, bacteria — that live on dead and decaying life take resources from each other. The same holds for plants, whose lifestyles seem superficially the most innocuous. One plant shading another is a parasite that takes sunlight, a vital resource, without giving anything back. All life needs resources and thus takes from other life, competes with other life, and kills other life. Thus all life is parasitic, for the good of one organism means the harm of another. And yet, even mortal enemies are ultimately linked in fate. This is the central paradox of self and other.

Although this paradoxical tension may now be obvious, it may be less obvious why many organisms seem to harm others unnecessarily. If parasites harm themselves by damaging the host, why have they not discovered that? They could do better for themselves by doing better for the other. Why don't they? For instance, what if a parasitic wasp could damage its caterpillar host so little that the caterpillar could survive and reproduce? In doing so the wasp would have served other and self alike. Why doesn't it?

On the surface, there are many reasons, perhaps as many reasons as ways to harm others.[42] Parasites may simply be shortsighted about their best interests.[43] (We humans are certainly in no position to point fingers at other organisms for shortsightedness.) Other parasites may carry the burden of a history that limits their future: they have come to rely fully on the host, having lost capabilities that would allow them to do what might be best for self and other.

But these superficial reasons may hide a more general principle: that harm is a price paid for the opportunity to create novelty. It is a price paid for new encounters and relationships between self and other, from the evolution of many-celled organisms to that of elaborate societies — whether those of insects or those of humans — and, ultimately, for the opportunity to influence the course an open world takes.

Related Prisoners

A view different from the biological view on self and other originated in economics and social psychology. It is worth exploring because it is similar to the biological view in one key respect: self and other are extremely distant yet are equally conjoined in it. Although conceived with people in mind, this view also sheds light on a wide range of animal behavior, from how female iguanas fight over egg-laying burrows to how male dungflies leave one cowpat in the quest to find another cowpat and, ultimately, a female.[44] This view is best illustrated with a game played countless times every day around the globe, a game whose outcome seals the fates of people, companies, governments, and nations.

The following imaginary situation contains the essence of this game. You and an accomplice, professional burglars, break into somebody's house and steal a lot of money. The police imprison you as suspects, although they cannot prove that you committed the crime. While the police investigate, you cannot communicate with your accomplice because you are both held in isolation. If both of you deny committing the robbery, you each might get away with a short prison sentence of one year on lesser charges, such as possession of illegal weapons, because the police might not be able to convict you. During the interrogation, however, the police offer you (and, independently, your accomplice) a deal.

The police ask you to betray your accomplice. If you do, and if he remained silent about your involvement, you would get away free and he would get the maximum sentence of six years. The opportunity seems tempting. You could perhaps even twist your story to make yourself look good. Everything was his idea, you might say, and he coerced me because I owed him money. But what if he had the same idea? If you betrayed him, and if he betrayed you, the police would give both of you a large sentence of four years. But what if you did not betray him, what if you remained silent but he did not? Then you alone would get the maximum sentence of six years, and he would walk away.

In sum, you would both get off with a light sentence if you sealed your lips, if you *cooperated* — without communicating. If you betrayed him and he stayed silent, you would go free and he would get six years. In this case, we might also say that you *defected* and that he cooperated. Conversely, if he defected and you did not, you would get six years and he would go free. And if you both defected, you would both get equally many years. This game is called the prisoner's dilemma game. It is a dilemma because self has to make a vital decision whose results depend on the other's decision, yet self cannot know the other's decision before it is too late.

This game applies — despite its name — to innumerable real-world situations in which two parties have conflicting interests and no information about the other's intentions. Both would benefit by cooperating. If one — the defector — betrays the other, the defector benefits, albeit less than if both cooperated. If both defect, then both do poorly.

The parties need not be two people. They might be organizations, such as businesses, armies, or governments. For example, two nations on the brink of war might benefit by maintaining peace. But if one tried to crush the other with a preemptive strike, it might gain even more, although at some cost of life or resources. And if both declared war, they might run each other into the ground in a lose-lose scenario. The dilemma arises because one side does not know the other's plans and resources. Many other situations — arms races, labor contract negotiations, legislative conflicts, and negotiations among companies — can be cast in terms of this game. Often there are complicating factors, however.

First, the game is usually played not once but many times, because negotiations — whether about divorce settlements or arms races — often involve many steps. Each step is one round of the game. And opponent and conflict usually do not disappear at the game's end, regardless of how the parties negotiated. The parties to the conflict — countries, competing businesses, or a divorced couple with shared custody of the kids — will have to coexist, perhaps indefinitely. Their future may depend critically on each step. Second, each round of the game will affect both the future of the players and the future of the negotiation. If you acted in bad faith previously, your opponent might retaliate. If you cooperated, the opponent might cooperate, too. Third, the negotiation is usually open-ended: it may take months, years, or even decades. For instance, when a company is losing money, negotiating a labor contract may be an excruciating back-and-forth process between unions and management. When ideological lines have hardened, arms reduction talks may become endless. And who does not know about political parties gridlocked in legislative negotiations — be they about health care, civil liberties, or defense spending.

Social psychologists are in the business of finding out how people behave in such games. They orchestrate experiments in which two people play a game once, a preset number of times, or an unknown number of times, as in a real-world conflict. They also want to know whether any one strategy is superior to other strategies, leading to greater overall benefit to the players. Robert Axelrod from the University of Michigan wanted to find the best strategies for the prisoner's dilemma game. To this end, he challenged other scientists — mostly game theorists — to a computer-based tournament of the game.[45]

Axelrod asked each participant to submit one strategy for playing the prisoner's dilemma game. A strategy is a mechanical rule for playing, such as "No matter what my opponent does in each round, I will defect"; "No matter what my opponent does, I will cooperate"; or the more complicated rule, "I will cooperate in the first round of the game, but if the opponent defects in one round, I will retaliate and defect in the next round. If, on the other hand, the opponent cooperates in one round, I will cooperate in the next round." Axelrod let a computer play these strategies against each other. He then asked which of the strategies is most successful in general. That is, when played against *any* other strategy, which one benefited its player the most.

One lesson from Axelrod's work is that human players will defect more often if they know how many rounds they will play than if the game is open-ended. And they are most likely to defect if only one round is played. But what if the game is played for an unknown number of rounds? It might end after ten rounds, ten thousand rounds, or never — thus becoming the game of one's life, ending with one's dying breath.

An almost universal best strategy in this situation is the last of the three I just mentioned. It is known as "tit-for-tat": whenever you cooperate in one round, I reward you by cooperating in the next. And if you defect, I defect in the next round.

Tit-for-tat turns out to be an excellent strategy, beating almost all other strategies in open-ended games. Inferior strategies include those where a player always cooperates. This strategy is vulnerable to nasty, defecting opponents. An equally inferior strategy is constant defection, because its players shoot themselves in the foot and because both players gain more if they both cooperate.

How does all this relate to our conversation about self and other? In this game, self and other are prisoners in their world, unable even to communicate. If the game is played by large institutions or nations, the people behind the decisions may not even know of one another's existence. Yet they are conjoined by their fates, for every decision by self will have great impact on other. The more rounds are played, the longer their relationship continues, the more that is true. A shrewd and nasty gamble against a one-time opponent might just pay off, whereas the same behavior against a long-time opponent may be fatal and ruin the future for both. In a long-term relationship, both are much better off cooperating and not defecting, steering away from a path of escalation, yes, even assisting the other.

Universal Relatedness

You might view the success of tit-for-tat as selfishness in disguise: "I do things for you, so that you will do things for me." But consider this success in light of a perspective that emphasizes not one player but the relationship between them. What could be more separate than the two players in the prisoner's dilemma, say, two hostile nations in a nuclear arms race? And what could be more linked than the fates of those two nations? The one launching its missiles first risks triggering a nuclear holocaust, turning the world into a radioactive wasteland, and obliterating itself together with all human life.

This perspective, emphasizing relatedness among self and other, is far from new. For example, few would accuse the following sources of advocating selfishness:

> What is hateful to you, do not to your fellow men. That is the entire Law; all the rest is commentary. — The Talmud

> This is the sum of duty: do naught to others which if done to thee would cause thee pain. — The Mahabharata

These statements are the endpoints of a journey through a vast territory. It has included cells sacrificing life to become part of multicellular organisms, parasites exploiting their host yet depending on its survival, organisms — genetically related or not — taking great risks for each other, and humans linked through decisions of global consequences. This journey was a high-altitude flyover, revealing only the hazy outline of the domain. The examples of this chapter are but a glimpse of the billions and billions of relationships between self and other that this planet harbors and that have shaped it since life began. Each relationship would illustrate the same basic principle: when self and other enter a relationship — however separate they may be — a common fate is created. In some cases common genes are responsible, but in many others they are not. And the closer we get to the human realm, the more dazzling the spectrum of possible relationships:

common nationality, language, religion, moral values, profession, culture, all these and many more can bind fates together. Some people may devote themselves completely to an organization, whether a company, a country, or a religion. Others may readily sacrifice their lives to a vision. And they may be satisfied if they live on in the memory, the values, or the well-being of others. Still others lead lives centered on their empathy for others. (To "empathize" — psychologists say — is to create a relationship between self and other.)

Are there limits of self's ability to relate to other? Consider the following perspective expressed in a Navajo chant.

> The mountains, I become part of it . . .
> The herbs, the fir tree, I become part of it.
> The morning mists, the clouds, the gathering waters,
> I become part of it.
> The wilderness, the dew drops, the pollen . . .
> I become part of it.

Humans' ability to perceive and create relationships between self and other is almost limitless.[46]

Universal Separation

Lest this uplifting view absorbs us completely, let me point to its principal blind spot: this view sees only the whole coin, not its two sides. It elevates the relationship between self and other above all else. It is just as limited as the view emphasizing one side, self or other.

If everything is related, why not get rid of self and other, or even of the notion that the world is made of separate objects? I was just reminded why: a moment ago, I stubbed my toes against my writing desk, hard. That hurt! Such a painful experience is a vivid reminder that self and other, even though related in fate, are also utterly separate.

Countless destructive relationships among the living illustrate this principle. We have encountered some of them already and will meet others later. They include parasitic insects inside a caterpillar, their crawling breakfast, lunch, and dinner box; tumors, made of renegade cells that selfishly divide and devour their host alive; honeybees that use stings to defend their kindred and gut themselves in the process; plagues wiping out millions and themselves with them. And what about humans? Not even a human committed to a principle of universal relatedness is exempt from destructive relationships. For all human life depends on killing other life, whether animal or plant life. Living is a murderous business.

Nothing illustrates this better than two nations in the rage of war, killing millions, destroying the work of generations. The opponents bitterly awake from their bloodthirsty rage, often incredulous about how that rage could have arisen. And often little but the burden of history sustains political feuds. Just think of decades-long conflicts in the twentieth century involving the two Koreas, or East and West Germany. Such conflicts throw millions into misery. They show that humans may be the true masters of taking a short-term view of self and other, of ignoring how self and other are related. For our purpose, they remind us that self and other, however closely related, are still as utterly separate as the two faces of a coin. This does not mean that we must continue to elevate self over other, a view that has dominated biological thought for a long time. The supremely difficult task is to balance these perspectives, to switch between them, to accept the tension between self and other, and to stand in the paradox of their sameness yet separateness.

And again, why do destructive relationships in which self and other damage and destroy one another exist? They are the price to pay for the opportunity of creation. If we look back in time, we see that many relationships between self and other—however tenuous and risky at first—evolved into marvels of creation. Neither of us would be here had not some parasitic cell about 1.5 billion years ago become part of another cell to provide energy; had not countless other cells given up life to become part of multicellular organisms; had not billions of other kinds of cells done the same while we differentiated from fertilized eggs into humans; had not parents sacrificed much to raise us, care for us, and teach us everything we know.

CHAPTER 3

Wholey Parts and
Partly Wholes

Inasmuch as you have done it unto the least

of my brethren, you have done it unto me.

Matthew 25:40

Have you ever watched a swimming *Escherichia coli* bacterium through
a microscope? If not, you would be impressed. For one thing, the bacterium looks very different from the dead stuff around it. In a sea of inert
and dead debris, from barely visible specks to hulking boulders, the bacterium is an island of frantic activity. It darts in one direction, stops,
tumbles around erratically, as if undecided where to go, then darts in
another direction, stops again, and so on. This bacterium — a swimming
whole — has billions of molecular *parts*.[1] Some of these molecules are absolutely essential for swimming.

Which is more important for swimming, part or whole? Is one of them
merely a mirage? Parts influence the whole, but can the whole also shape
its parts? Swimming bacteria, shape-shifting amoebas, and carnivorous
colonies of marine organisms are among the examples that illuminate the
relationship between part and whole in this chapter.

The question how parts and wholes are related is not just academic:
answering it helps us choose our place in the world and shows us how we
can shape this world through our actions. For example, it helps us learn
whether society (a bigger whole) or our genes (smaller parts) run our

lives and to what extent we can control these lives. The examples below will illustrate some general principles. Paramount among them is that to answer this question requires *choice*.

Shapely Wholes

In chapter 1, I described how cells converse using molecular messages and their receptor proteins. These and other proteins are a good place to start in studying wholes and their parts.

Proteins consist of smaller building blocks, amino acids. Each amino acid has two parts akin to docking sites.[2] By means of each site, an amino acid can be linked to another amino acid. Much like two links in a chain, two amino acids connected through these sites are firmly joined. To either link, a third amino acid can attach via its docking site. And another. In this way, long chains of hundreds or thousands of amino acids can be built. Proteins are such chains.

Proteins contain twenty kinds of amino acids that differ in shape. Chains of these amino acids can twist, bend, and fold into complex and sometimes bizarre three-dimensional shapes. Their flexibility derives from the amino acids themselves, for much like adjacent chain links, two amino acids can bend and twist. The resulting flexibility and three-dimensional folding are essential for how proteins work. Receptor proteins, for instance, cannot read molecular messages unless they are folded into a three-dimensional shape.[3]

Some proteins fold into ball-like and others into rodlike shapes. Yet others fold into shapes that look like barrels, pretzels, or fists. Regardless of a protein's overall shape, some amino acids are hidden inside the protein, while many others form the protein's surface. This surface may look very different depending on which of twenty possible amino acids occur on it. But it is always richly textured, covered with high peaks and gently sloped hills separated by valleys that can be deep ravines or shallow dips. The shapes of the surface amino acids create a protein's rugged relief, its big shape made of smaller shapes, the complex face it presents to the world.[4] These shapes — millions of them in every living cell — sustain life as we know it. They make molecular conversations possible, promote chemical reactions, provide a support scaffold for the cell, and transport other molecules.

For proteins, the relationship of part to whole is easier to grasp than for other parts and wholes, for example humans and their societies. This is because the relationship between a protein's amino acid parts — a flexible molecular hinge — is much simpler. And here is a key question about these protein wholes, a question whose answer will take us a long way toward grasping the relation of part and whole in general. This question is usually asked about wholes other than proteins and has eluded biologists for decades and philosophers for centuries. Where is a protein's function located? Which of its amino acids are responsible for that function?

I will ask this question for one kind of protein, enzymes, which facilitate the chemical reactions that sustain all life. With enzymes, organisms can form the molecular bricks that facilitate growth and transform food or light into useful forms of energy. An enzyme binds one or more molecules and, by changing its shape, promotes a chemical reaction between them. That is, it joins, breaks, or rearranges the molecules' parts. These molecules are often *much* smaller than the protein itself. They must attach to the protein surface, near one another and in precise juxtaposition. The protein surface — shaped by the surface amino acids — determines where they attach. Being very small, these molecules cover only one or a few key amino acids that must be near each other to create the matching surface. These key amino acids, however, need not be neighbors on the string of amino acids. Think of this string as a chain of unequally shaped links rolled up into a ball. Two links of this chain may abut each other on the surface. Yet they may be far apart on the chain.

If the shape and position of a few surface amino acids determine where small molecules bind, are these few amino acids all-important for the protein's function? Are they the answer to our question? Yes, it would seem. But how do these key amino acids assume their position in space? What positions them to position other molecules? When you roll a chain of links into a ball, only a small part of the chain, a few links, will form the surface. The same holds for proteins: most amino acids may be buried in a protein's interior. Does any one of these interior amino acids make a difference to protein function? It seems as though it should not, because only the surface amino acids matter. But imagine unwinding the ball of links, taking a piece out, splicing the remainder back together, and rolling it back up exactly the same way. Would two previously adjacent surface amino acids still be next to each other?

Molecular biologists have been asking this question by engineering proteins with removed or altered amino acids. Such engineering may change a protein's shape. It may fold differently, and the amino acids juxtaposing two small molecules may become separated or move into the interior of the ball. It seems that the answer to our question has just become less straightforward.

Enzymes are not exceptional proteins in this regard. The same holds for proteins that give cells support and shape. Many of these proteins assemble into long rods that work a bit like tent poles to prevent a cell from collapsing. Each rod is a pillar made of protein parts. As with bricks in a pillar, most of a protein's surface will be embedded in the pillar, surrounded by other bricks. Change the surface of one brick, perhaps by only one amino acid, and the resulting pillar may be crooked and unstable or, worse, may not form at all. And change an amino acid in the protein's interior, and the effects may ripple upward to the surface. It may alter how the protein folds and thus distort the brick's surface. All parts of this whole may thus be important for it to function as a molecular brick.[5]

Reducing a protein's function to individual amino acids may thus be impossible. Perhaps, then, we should consider that only the overall shape of the whole protein matters. Whether a brick is made of clay, concrete, or plastic may not be important, as long as the brick is rectangular. In the world of enzymes, this possibility translates into a concrete question. Are there molecules completely different from proteins, molecules with similar shapes that have completely different parts, but that — like enzymes — enable other molecules to react? Yes — ribozymes, which are ribonucleic acids (RNAs) that enable other molecules to react, much like protein enzymes.[6] Their building blocks, however, are ribonucleotides, chemically very different from the amino acids of proteins and very similar to the deoxynucleotide building blocks of our DNA.

Armed with such an example, we might now rush to embrace the opposite extreme: forget about the parts; really it is the whole that counts. Only overall shape matters. But beware! Pick one particular molecule — whether protein or RNA — that enables two other molecules to react. Just one specific molecule. Then change its parts; you will have changed the whole and thus its function. The whole depends on its parts, whenever you are dealing with a *concrete* object. (And if you look up from this book for a moment, you see that concrete objects are what the world consists of.)

You could spin all this further, heaping argument on argument for either perspective, that emphasizing part and that emphasizing whole in explaining the properties of proteins or any other object. In fact, for some wholes — the human body and its parts or societies and their people — this string of arguments has been spinning for hundreds of years. And in all these centuries, the question whether part or whole is more important has not been definitively answered.

This failure has a reason. For if you insist that either part or whole must be primary, you are trapped in a vicious circle — ultimately, a paradox — such as those we encountered before. Consider rather that part and whole are two sides of the same coin, completely separate yet inseparable. Changing the whole means changing its parts; changing its parts means changing the whole. Part and whole have this particular relationship, and if you account for it, the question where a protein's function is located has no definitive answer. Of course, to take this perspective, one has to be willing to give up the primacy of either part or whole and to live with the tension that their relationship entails. This tension can be resolved only by focusing on one of the two opposites, however limited and ultimately wrong that choice may be. The examples in this chapter will illustrate this tension.

The Smallest of Parts

We often tacitly assume that smaller parts are more removed from the whole than bigger parts; that the smallest parts cannot be equal partners in shaping the whole; that they are not simply the other side of the coin.[7] For example, a person that is part of a family or community might readily influence that whole; but what about a cell inside this person, or a protein inside the cell? Can they also influence the whole?

To explore this question, I will use our enzyme example and break it into its smallest parts. For variety, imagine the whole formed by an enzyme and two small molecules, A and B, which the enzyme joins. How do changes in this whole affect the chemical reaction of A and B? A, B, and the protein's amino acid parts consist of even smaller parts, atoms such as nitrogen, oxygen, and carbon. An amino acid's features (its shape, electric charge, and ability to connect to other amino acids) depend on these atomic parts. So do the properties of the molecules A and B. These atomic parts in turn have parts, which are elementary particles such as the protons and neutrons forming an atom's nucleus and the electrons forming its shell. The number of protons, neutrons, and electrons give an atom its identity, including mass, size, stability, and the chemical bonds it can form.

I have already mentioned that eliminating parts of an amino acid string can change a protein's shape. But what about changes on a smaller scale that alter only one amino acid—perhaps by as little as one atom? Can they affect the whole just the same? Yes. The fundamental reason is that all parts of a protein, from amino acid down to the smallest atoms, take up space. Replacing one part with a larger part pushes other amino acids aside and may force the protein to fold differently. Replacing one part with a smaller part may create empty space into which another amino acid might squeeze. The protein can then fold into a smaller shape with changed surface. A similar argument also applies to the small molecules A and B: a change in either can destroy their match to the protein's surface and thus the protein's ability to join them.

The smallest conceivable changes in the electrons, protons, or neutrons that make up individual atoms are no exceptions. Electrons or protons carry negative and positive electric charge. And to change these particles is to change—at the least—an atom's charge. Such change can have drastic effects: after such a change, an enzyme that previously attracted a negatively charged small molecule may now repel this molecule and prevent a chemical reaction.

Neutrons, however, do not carry charge. This argument therefore cannot apply to them. To add or eliminate a neutron to an atom is to generate an *isotope* of that atom, a version of the atom with different mass. When this isotope is incorporated into a molecule, the molecule would simply change its mass by a small amount, less than one part in a million for large molecules like enzymes.

Even such a small change in mass can make a difference. The difference it makes is best illustrated with our enzyme that joins two molecules, A and B. Imagine that the enzyme has a choice between two forms of A, one heavier than the other because one of its atoms is a heavy isotope. You could compare the reaction joining A and B to the process of suspending a door in its frame. Have you ever hung a door? You would lift the door— perhaps with difficulty—and then lower it carefully onto its hinge. Doing so takes energy. And so may the joining of A and B. If A has become heavier, even by as little as one neutron, the protein needs more energy—

as if it had to hang a heavier door. And sometimes there may not be enough energy to connect heavy A to B. If so, a protein that could have joined the light isotope A to B leaves the heavy isotope A untouched.

This is an example of a phenomenon called isotope fractionation: enzymes prefer some isotopes over others, despite minute mass differences between them.[8] Isotope fractionation influences the composition of living tissues, because the enzymes that build these tissues prefer to use building blocks with certain isotopes. In sum, changes in parts of molecules as small as neutrons can influence the atomic composition of whole organisms. Let this one example stand for innumerable others: an unbroken chain connects the smallest parts to the largest wholes, from atoms to organisms and even further. Even the smallest conceivable parts of a whole can influence the whole.[9]

Conversely, wholes can influence their parts, down to their smallest parts. Proteins, for example, not only can transform other molecules but can transform themselves. Some proteins attach a few extra atoms to one of their amino acids. Other proteins can self-destruct by chopping themselves into pieces. For more familiar wholes, this is so obviously part of your everyday life that you do not even think about it: your government requires you to pay taxes; you may have served involuntarily in the military; you appear at work on time every morning because your company regulations say so; all kinds of social rules make you behave in peculiar ways that might be incomprehensible to an outside observer, say from a different culture or planet.

The Parity of Part and Whole

What all this means is that the relationship of part and whole is mutual, like that of the two sides of a coin. They are separate yet inseparable, equal partners in stability, destruction, or creation. This mutual relation of parts and wholes — one shaping the other — holds for all wholes and their parts, not just for enzymes or human societies. Even the most disparate kinds of objects may obey this relation.

Perhaps the most dramatic illustration comes from physics.[10] It speaks to the relationship of a particle physicist and her apparatus — the whole — to the smallest parts of matter she is studying through experiments. A fundamental tenet of quantum physics holds that the very small has a dual nature. It may appear as wave or particle, depending on the question an experimenter, the physicist with her apparatus, asks of it. In this sense, the whole can influence the nature of the smallest things in the world. Conversely, how the particle behaves in such an experiment influences the whole. It influences the state of the apparatus, the cacophony of communication crisscrossing the physicist's brain, and perhaps her perspective on the world (and maybe even that of other scientists).[11]

Swimming Wholes

We will now enter the territory of living wholes. To explain what life is may be difficult. But how to recognize it becomes crystal clear with a mere glimpse at a bacterium that darts this way and that under a microscope. Less clear is the purpose of the bacterium's frantic activity. A swimming bacterium reminds me more of my erratic trajectory as I scour my office for lost objects — keys, glasses, or wallet — than of a clever little organism. But there is method to a bacterium's swimming, perhaps more than to my haphazard search. What causes a bacterium to move? The answers to this question will speak volumes about the relationship between whole and part.[12]

The gut bacterium *Escherichia coli* and many other bacteria swim with flagella that resemble thin curly hairs or corkscrews.[13] Viewed up close, flagella are stiff protein rods, up to ten per cell. They twirl blazingly fast and push the bacterium through the water like a ship's propeller. But although the most important part of going places is going forward, it is not everything. Imagine a car that can go only forward, straight as an arrow. It could go no farther than the next bend or the end of the road. Cars can steer, but — surprisingly — bacteria cannot. They can only move forward. So how do they ever get anywhere?

Flagella can rotate clockwise and counterclockwise. Only one of these directions, counterclockwise, pushes the bacterium forward. When rotating counterclockwise, all the flagella on the cell line up in one direction and jointly propel the cell forward. You might think that the opposite, clockwise rotation, would do exactly the reverse: pull the cell backward. Not so! Clockwise rotation does something entirely different. When its flagella turn clockwise, the bacterium tumbles around as if undecided where to go. It stays in place, turning repeatedly and erratically. The flagella point not in one but in many directions, and because each pushes in its own direction, the bacterium goes nowhere.

A swimming bacterium switches rapidly between these two directions of rotation. With flagella rotating counterclockwise, it moves straight ahead. Then it stops, the flagella start to turn clockwise, and the bacterium tumbles. Then they switch direction again, and the bacterium resumes a straight path. But because it has tumbled and thereby randomly reoriented itself, it now moves into a different direction. Most bacteria know only these two ways of moving, tumbling in place and going straight ahead. This is a step up from just going straight, but it may not help clarify how a bacterium ever gets anywhere. To find the missing piece of information, we must ask why bacteria move around to begin with.

Bacteria need food to provide them with energy and building material to grow and divide. Most bacteria can survive on much simpler foods than we can. These foods consist only of a few kinds of molecules: sugars like sucrose, alcohols like glycerol, or acids like citric acid. But once a bacterium stays in a place for a while, it depletes all the food at that place, and then the bacterium has to go looking for more. Bacteria are very sensitive in detecting food in their surroundings, and as the bacterium rests in one location, it may detect a few food molecules drifting by, a faint waft from a distant kitchen. But where is the kitchen?

Imagine you are hungry, stranded somewhere in the wilderness, in the middle of a dark night, far from the nearest settlement. Suddenly you smell food, faint but delicious. Somebody is roasting a kill over a campfire. But neither sounds and sights nor wind help pinpoint the source. Your eyes and ears are of no use. (Bacteria do not even have any.) What would you do? You would try to find the direction in which the smell gets stronger. Lacking any other cues, you might start moving into one direction — any direction. If the smell gets stronger, you keep moving. If it gets weaker, you change direction.

Bacteria do exactly that. When a bacterium senses a waft of food, it starts to move. It moves straight ahead, but in an arbitrary direction. There are two possibilities: it is moving in the right direction, approaching the food, or heading in the wrong direction, away from the food. The direction is right if the smell is getting stronger, and the bacterium keeps moving straight. If it is moving in the wrong direction, the smell gets weaker. But the bacterium cannot simply steer away from this direction. It reverses the flagellum's rotation, reorients itself by tumbling, reverses rotation again, and resumes moving straight ahead. The new direction may be completely different from the one before. It may be the right direction, in which case the bacterium stays on course. It may be another wrong direction, in which case the bacterium tumbles and start again. Eventually it finds the food source.[14]

Getting to food in this way may seem inefficient. But lacking other cues, you might do much the same. It may in fact be the only possible thing to do. And keep in mind that you, as a lost traveler on solid ground, have just four directions to choose among. A swimming bacterium has a more difficult job, because it also needs to choose between swimming up or down.

The bacterium's swimming for a living is one of the simplest purposeful behaviors we know.[15] It serves as a metaphor for purposeful behavior in general, which is a hallmark of the living. I mean any purposeful behavior, from seeking food, mates, shelter, or work to very complex behaviors such as negotiating an arms reduction agreement, building a company, or running for elected office. All these behaviors have a goal, and getting to this goal requires choices.[16] With this commonality in mind, one may ask what — if anything — in the bacterium "chooses" to change swimming direction? And are any of the bacterium's parts more important than others? To answer these questions, we need to delve into further details of bacterial swimming.

Molecular Marvels

A bacterial flagellum is driven by a molecular motor, a sophisticated contraption made of protein molecules. Much like an electromotor, this motor consists of a stator and a rotor. The stator anchors the flagellum to the cell wall. The rotor turns the flagellum. This motor is impressive: its force propels the cell up to ten times its body length per second; it is extremely efficient, tiny compared to the huge flagellum, but enormously powerful, turning at up to 15,000 rpm and able to change directions nearly instantly, within half a turn.

To see how and when the motor changes directions, we need to know how bacteria detect food molecules. They do so through protein receptors in their cell walls that bind food molecules. The principle here is very similar to our earlier examples of chemical conversations—involving receptors and messages—that shape organisms like you and me. When a food molecule drifts by a cell and bumps into a receptor, it attaches to the receptor, which changes shape. (After some time, the food molecule may drift onward, and the receptor may revert to its old shape.) Thousands of receptors for food molecules populate a bacterium's cell wall. As a cell approaches a food source, more and more food molecules bump into these receptors, and more and more of the receptors change shape. Exactly the opposite happens when a cell moves away from the source: more and more receptors revert to their original shape.

What happens as a receptor changes shape? As in our earlier example, a protein inside the cell detects this shape change. This protein is called CheA.[17] Think of CheA as having a surface matching the changed receptor. It thus recognizes this change and alters its own shape in response. In the absence of food molecules, CheA is a peculiar kind of enzyme, because it does not allow two small molecules to react but instead attaches a small molecule—a phosphate—to another protein called CheY. Let's call the modified CheY protein, after CheA has attached a phosphate to it, the CheY-P protein (the "P" stands for "phosphate").

After CheA has changed shape in response to a receptor's encounter with food, CheA can no longer transform CheY into CheY-P. This is crucial, because CheY-P is an important link in the chain from food to swimming: it floats freely through the cell and thus occasionally bumps into the motor driving the flagellum. This encounter has a peculiar effect on the motor. If the motor turns counterclockwise, propelling the bacterium forward, CheY-P makes it switch into the clockwise, tumbling direction. Again, you can think of this in terms of shape change, where CheY-P has a shape precisely matching that of a gear lever on the motor. When CheY-P bumps into the gear lever, the gear lever shifts into reverse.

If the bacterium swims toward a food source, more and more food molecules arrive at the cell wall; more and more of the receptors change shape; the cell contains many CheA proteins, and an increasing number of them change shape as well; as a result they can no longer make CheY-P from the CheY proteins in the cell; and thus the motor does not change direction.[18] The bacterium keeps swimming straight toward food. Conversely, if the bacterium swims in the wrong direction, fewer and fewer receptors change shape; CheA proteins are free to make CheY-P; CheY-P molecules bump into the motor's gear lever; the motor reverses direction; and the cell begins to tumble. After some time, CheY-P molecules detach from the motor and the motor reverses again to straight swimming. This conversation of shapes, of countless colliding molecules—food, receptors, CheA, CheY, motor proteins—takes place nearly simultaneously. That is, it occurs so rapidly that a bacterium can change directions several times per second.

Everything I have said previously about wholes and parts applies here as well. Part and whole, however chosen, are like two sides of a coin, utterly separate yet inseparable. They mutually influence each other. The amount of CheY-P in a cell influences where the bacterium is going to swim (as do all other parts of the signal chain). And whether the bacterium swims the right or the wrong way influences how much CheY-P a cell contains.

We can now revisit the key question: What makes a bacterium move? And, more precisely, what causes the bacterium to change directions? From what I have said, you might argue that CheY-P must be the key part. For only if CheY-P bumps into the motor does the bacterium start to tumble. But what about CheA? CheA, after all, makes CheY-P. (No CheA, no switching directions.) Alternatively, you might argue that the receptor must be the key part. Unless it changes shape, CheA will not make CheY-P. But then we must not forget the food molecules. Unless they bump into the receptor, nothing will happen. And finally, there is the motor. Without a gearbox, it would incessantly turn one way.

All these answers focus on individual parts. But many other answers do not involve these parts at all. For example, one could argue that all these molecules together and, in this sense, the bacterium as a whole switches direction. Or we could move beyond this whole to even larger wholes. Suppose that other bacteria have discovered the food source before our bacterium. While our bacterium gropes its way toward the food, these other bacteria devour most of it. As our bacterium approaches what used to be a food source, the waft of food gets weaker instead of stronger. Eventually, our bacterium swims elsewhere. Here, you could argue, the other bacteria make our bacterium change directions. And last, imagine you are a scientist studying what makes bacteria move. You ask a bacterium a question by putting food near it and notice that it swims toward the food. Then you remove the food source, and the bacterium mostly tumbles, not knowing where to go. *You* caused that change.[19]

These answers to our key question are very different. None of them is the one *true* answer, because each one emerges from one of many valid perspectives. Which perspective you choose will depend on the scientific conversation you want to have. To study the role of external molecules in swimming, you might deposit such molecules in a bacterium's surroundings. You would then find that some molecules cause direction change, whereas others do not. To study the role of CheY-P, you might compare cells that contain CheY-P with others that do not. You would then find that CheY-P is responsible for direction change. To focus on how bacteria interact, you would ask how other bacteria influence a bacterium's swimming direction. All these perspectives are valid choices. They are, however, meaningful only in a specific conversation you choose to have and as answers to specific questions.[20]

Swimming Genes

It may be easy to see that swimming bacteria offer a variety of choices to delineate parts and whole. But elsewhere, we often have trouble seeing such alternatives as valid. More often than not, we ferociously defend *one* perspective on whole versus part as the truth.

This is true especially when the stakes are high, where our choices can influence human lives. Consider the following question: What is responsible for intelligence, a few genes, all our genes, or the world around us — embodied in our parents, our teachers, our nutrition, or our income? Consider that this question is completely analogous to asking what causes bacteria to swim in a particular direction: a protein or other bacteria. The temptation of clinging to one perspective here comes from what is at stake. If you are a policy maker (or if you help elect policy makers), then your choice of perspective on this question might affect the lives of thousands or millions of people. Should we invest in better education for the disadvantaged or write them off as a lost cause? Should we alleviate income inequality or let the rich get richer? Should we improve the health of all people or focus our resources on those who can pay? The perspectives we take on these and innumerable issues depend on whether we think that some people are uneducated, poor, or sick because of their poor genes, which caused poor intelligence and bad choices, and that we can do nothing about it.[21]

Genes have not yet figured prominently in our conversation about behaviors like bacterial swimming or human intelligence. Let us examine their role in bacterial swimming more closely. And consider that everything I say would apply to murkier subjects such as human altruism and intelligence.

The entire signaling chain for detecting and communicating food-related information—from receptor, CheA, and CheY to the motor—consists of proteins. I have already mentioned where proteins come from: the cell makes them from information contained in genes. Here's how.

A gene consists of a sequence of four nucleotides, abbreviated by the letters A, C, G, and T. Proteins, on the other hand, consist of a sequence of twenty amino acids. In making a protein, a cell translates a gene's letters, in groups of three, into the chain of amino acids. Consider part of a gene that reads as follows: ATGGGA . . . (this sequence might continue for thousands of more letters). The cell takes the first three letters, ATG, and translates them into one amino acid called methionine. Then it reads the second three letters, GGA, and translates them into another amino acid called glycine. That is, the cell builds a chain of amino acids, a protein, in which a methionine is followed by a glycine. As the cell reads further, it encounters the next group of three nucleotides, interprets them as another amino acid—let's call it X—and appends X to the growing protein chain, making methionine-glycine-X. This process continues to repeat, with each group of three nucleotides designating a particular amino acid.

A bacterium contains a gene for each of the proteins involved in swimming, including the receptor proteins, CheA, and CheY.[22] But I said that a bacterium contains many receptors and many CheA and CheY proteins. Does it then contain as many genes as, say, receptor proteins? No, it contains only one. To make many receptor proteins, the cell translates the information in this one gene many times.

How do genes influence the bacterium's (and our) behavior? To answer this question, it is best to compare two organisms. Let's pick two bacteria from a population and focus on one protein, say a part of the motor, and the gene responsible for making it. The two bacteria might have slightly different forms—alleles—of this gene. The DNA text of these alleles might differ by as little as one letter. Even this small difference could cause an exchange of one amino acid for another in the motor protein, rendering its shape different in the two bacteria. One of these motor proteins might cause the motor to turn more slowly, or — if part of the gearbox — it might recognize the switching signal CheY-P more or less easily. (Think of different copies of a key, one fitting smoothly, whereas the other requires fiddling.) These slight differences may affect the bacterium's life enormously. For instance, impaired by a slow motor or a motor whose gear lever sticks, a bacterium may run out of fuel before getting to food.

Genes influence behavior in this sense. Their different alleles cause bacteria to swim differently. Geneticists have used this observation to dissect the cell's parts important for bacterial swimming. To identify the role of a cell part, geneticists change a gene, for example, the gene for a motor protein. "How does this change affect the motor?" they then ask, and from the bacterium's response they guess the gene's role. This approach is so powerful that virtually everything we know about the relationship between genes and organisms derives from it.

As always, powerful perspectives have dangerous blind spots. Where is the blind spot here? We already encountered it. It is the same blind spot that would make you defend CheY-P (or a receptor or a food molecule) as the *one* part that causes the bacterium to change directions. Keep in mind that the motor is only partly responsible for swimming and that other proteins — each with its own gene — are equally important. Two swimming bacteria may contain different alleles of the receptor gene, say a "fast" and a "slow" one. The protein translated from the fast allele might match a food molecule better, or it might change shape more rapidly. Furthermore, two swimming bacteria may contain different alleles — "fast" and "slow" — of the CheA gene, one recognizing receptor shape more readily than the other. And exactly the same idea holds for different alleles for the CheY gene.

Imagine that you pick two bacteria that swim in the same way. That is, they typically take the same amount of time to reach a food source. One of them might have a slow receptor allele, fast CheA and CheY alleles, but a motor that is slow to change direction when hit by CheY-P. The other bacterium might have a fast receptor allele, slower CheA and CheY alleles, but an allele for a very sensitive and fast gear lever in the motor. Yet overall they may swim equally well, their relative strengths and weaknesses neutralizing each other.

Now consider two bacteria where one swims better than the other. One may have alleles for a slow receptor, fast CheA and CheY, and a slow motor, but overall it ends up first at the food source. The other one may produce a fast receptor, slow CheA and CheY, and a faster motor but ends up second at the food source. Which of these genes makes the winning bacterium win? The one making a slower motor? That would be surprising, wouldn't it? The one making a fast receptor? Perhaps, because it seems that a fast receptor would be a good thing. (But even a bacterium with a fast receptor may end up second at the food source.) Thus, once we abandon a narrow focus on one gene, the problem of assigning responsibility for behavior to one gene has no *one* true solution.[23]

What I said before about the proteins involved in swimming holds just the same for genes.[24] The perspective that one individual gene is responsible for the behavior of the whole is perfectly valid. But so is a perspective focusing on any other gene. And each such perspective is limited, as can be seen from different gene combinations that accomplish the same. It is as limited as a perspective arguing that one particular protein is the one responsible for swimming. This limitation may be much easier to accept for swimming bacteria than for human behaviors, including intelligent behavior, risk-taking or violent behaviors, and many others.

This limitation leads us back to the perspective that whole and parts are like two sides of one coin: they are completely separate, yet we cannot understand any living thing by examining just one of them. They are also inseparably linked. To take this perspective is to admit a paradoxical tension between part and whole. Take any part of a whole. From one perspective, it may explain what the whole does. From another perspective, it is utterly insufficient. This tension suspends our preconception that there must be one *true* way to view any object and understand it.

But what would we gain by admitting that this tension is fundamental? First, we could better attack solvable problems by casting unsolvable ones aside. Most who think about parts and wholes belong to two camps, "holists" emphasizing wholes and "reductionists" emphasizing parts. If holists accepted this tension, they might concede the parts' importance and stop invalidating scientists' efforts to understand them. On the other hand, those emphasizing the primacy of parts — many of them scientists —

might concede the validity of other choices. If we stopped arguing for one true view and conceded that we had a choice in this matter, we might gain the ability to choose flexibly. We could then more easily choose perspectives that allow us to solve problems we humans care about, problems that require delicate decisions, such as how to best distribute resources in a country, how to provide for and educate a population, and how to improve its intelligence.

Brainless Intelligence

The paragraph above touched on our ability to separate the world into different wholes and parts, which raises the next question: When do we call a whole a whole?

For a swimming bacterium the boundaries of whole and parts may seem clear. But is this the case for all wholes? When do we call a cell a cell, an organism an organism, a community a community, or a species a species? For variety, the following examples explore this question with wholes larger than organisms. But their main lesson — that our choices are important for the answer — applies to anything, anything at all.

Slime molds are organisms with a curious lifestyle.[25] Most of the time they grow and divide as single-celled amoebae. Like other amoebae, a slime mold's amoebae unceasingly shift shape, extending stubby tentacles in front while retracting them at the rear. This is how they move, for the same reason bacteria do: to eat. While moving, they engulf food particles and digest them. But what if there is not enough food around? Do they shape-shift in place until they die? Do they rest, waiting for better times?

Something altogether different happens. In a population of starving amoebae, hundreds or thousands of them in an area smaller than the palm of your hand, some amoebae begin to release a molecule, a chemical signal, into their surroundings. As time goes on and food grows scarcer and scarcer, more and more starving amoebae release the same signal. Depending on where amoebae first began to starve, some location may contain a little more of this molecule. The amoebae follow this smell, crawling toward areas where it is most intense, while releasing more and more of the same signal. (The amoebae's response to the signal — crawling toward it — requires a chemical conversation with receptors.)

As many as a hundred thousand amoebae thus end up crawling toward each other and, eventually, all over each other. What a mess, you might think. What could *that* possibly be good for? But there is reason behind their behavior. The writhing pile of amoebae now transforms itself into a slug, a living, slithering mass of cells that can crawl in one direction, away from famine and toward feast. (This slug is much simpler than the slugs feasting on lettuce in a vegetable garden. It has no muscles, skeleton, or nervous system. Its ability to coordinate its parts is thus, in and by itself, a remarkable feat.)

After a while, having perhaps crawled farther than any one amoeba could, the slug stops. It lifts one end off the ground until it stands upright. Its high end swells repeatedly, until it becomes a thick round bulb held up by the rest of the slug, which has now become a slender stalk. Inside the bulb, innumerable amoebae change shape one final time. They transform into spores, durable and light cells that can survive starvation and dehydration and that wind or water can carry to faraway places. Stalk and bulb together are called a fruiting body. Eventually the bulb bursts, scattering the spores to the air currents and to their destiny. The stalk cells die. When landing near a food source, a spore metamorphoses into an amoeba that grows and divides — alone.

In sum, when life turns bad, single cells of a slime mold team up to crawl someplace better and spread some of their kind through spores. This lifestyle raises many intriguing questions. Do slime mold amoebae compete and race to the fruiting body's tip to be spared a certain death in the stalk? And what benefit could be big enough for stalk cells to give up life? These questions speak to other facets of our topic. For now, just consider this puzzle: Which is the organism, the amoeba or the slug?

Carnivorous Forests

Another example raises the same question.[26] Colonial hydrozoa are water-living animals, close relatives of jellyfish and sea anemones. Many hydrozoan species spend their lives as polyps, attached to a solid surface, much like the more familiar sea anemones. Each polyp looks like a strangely beautiful submerged tree, as little as half a millimeter tall. Its muscular hollow stalk is firmly anchored to the ground and widens in its upper part to surround a hollow space. This hollow is a stomach, because the tree is carnivorous. Its branches are tentacles surrounding a mouth linked to the stomach. Swaying seductively in any water current, the tentacles will wrap themselves around any small animal unfortunate enough to swim by. Thousands of cells on the tentacles' surface shoot deadly chemicals into this prey animal, killing it swiftly. Eventually, the polyp ingests the prey and transforms it, within hours, into a gooey, nutritious liquid. From this liquid the polyp derives energy for growth and reproduction.

Polyps in colonial hydrozoa can reproduce in different ways. Some species have male and female polyps that release sperm and egg cells into the water. Fertilization initiates a new generation whose individuals swim freely for some time before settling down. In other hydrozoan species, well-fed polyps form tiny hollow, stubby outgrowths at their base, two or three at a time. These stubs grow longer and turn into hollow tubes that creep along the ground like shallow roots. In time, little buds form at their tips. Each bud grows bigger and eventually turns into a polyp, complete with stalk, stomach, mouth, and tentacles. New and old polyps remain linked through these hollow tubes, which connect their stomachs. Once a new polyp has grown large enough, this cycle repeats. A colony of more than a hundred polyps, linked to each other through a network of hollow channels, can form in a matter of weeks.

Some hydrozoan species mix these two ways of reproducing. They form colonies with two types of polyps. The first type is the predatory or feeding polyp I just described. The second type is a reproductive polyp that has neither tentacles nor stomach. Its sole purpose in life is to release sperm or egg cells into the water. Once a sperm has found an egg, a new polyp forms, and it founds a new colony. The colony grows and grows, but in time, all polyps in the colony die. They live on only through their descendants, offspring colonies conceived through sperm and eggs.

What is the whole here, polyp or colony? To speak of colony instead of an organism already insinuates an answer: the polyp must be the organism. After all, it is a reasonably independent unit, especially in the hydrozoan species that do not even form colonies.

But this is not the only possible answer. You could also call the colony an organism. You could argue that an organism must consist of different parts that cannot survive alone and that share the labor of helping the whole survive or reproduce. These parts may be as different as the proteins that allow a bacterium to swim and the masses of tissue that form our internal organs. The reproductive polyps certainly fill that bill. Unable to feed, they cannot survive alone. And they certainly contribute to the colony's reproduction. But what about the feeding polyps? They can survive on their own. How do they contribute, if at all?

Imagine the following scene. It is a slow feeding day, and only one feeding polyp in the colony has captured prey. It could keep the food, using it to grow bigger. This might allow it to capture even more food, until it became the tallest tree in this forest. But this polyp cannot reproduce. And as it grows, the other feeding polyps would starve. Eventually this sterile bodybuilder would perish. Unknowingly, it would effectively be dead as soon as the reproductive polyps die, because they are its only means of reproducing.

Hoarding food is thus a shortsighted strategy. So what does a successful feeding polyp do? It turns into a food pump, contracting its muscular stomach rhythmically to force the liquefied prey through its stalk and, by means of the connecting tubes, into the stomachs of neighboring polyps. As soon as they sense the food, the neighbors also start pumping, distributing the food in turn to their neighbors. Eventually, all polyps in the colony contract rhythmically. This symphony of pumping stomachs circulates food that serves feeding polyps and reproductive polyps alike.

Neither solitary polyps nor colonies have anything remotely resembling a brain. Polyps have a few nerve cells that help them capture prey. That's about it. They cannot coordinate their action through a nervous system. Each polyp lives very much in its own world. And yet, it is able to act on behalf of the colony.

If you take a perspective in which a whole requires codependent specialized parts — like our organs — then a slime mold's slug and our colony of polyps qualify as organisms. But is that the one true answer to our question what the whole *is*? No, for you could choose entirely different perspectives. For instance, you could reasonably require that a whole remain distinct from other wholes, that it retain separateness and identity. That might apply to wholes like bacteria, you, or me. But it does not apply to a slime mold's amoebae. And it does not apply to colonies of polyps, because sometimes two colonies that come into contact while growing and expanding their boundaries can fuse and become one.

Another perspective might require that a *real* whole's parts must not compete at the expense of the whole. The amoebae of a slime mold would not qualify. As individuals, they compete fiercely for food, and they jockey for the top position in the fruiting body. In contrast, it would seem that our polyps fill that bill. But these polyps can also compete with unexpected fierceness.

A miniature diver gliding through a hydrozoan colony would see a forest with a closed canopy of swaying tentacles. Tentacles of neighboring polyps almost touch. Whenever one polyp embraces its prey, neighboring polyps smell the prey's presence. They can flex their muscular stalks and bend over, wrapping their tentacles around the prey as well, and try to snatch it away from the other polyp. Thus can begin a tug of war that lasts for hours before one of the polyps wins and takes the prize home. In this situation it clearly might help to have a brain, because as soon as the winning polyp has liquefied its prey, it will serve the nutrients to its competing neighbor.[27] Does this competition mean that the polyp is the organism rather than the colony?

Many other examples illustrate that demarcating a whole requires choice. Some are quite familiar, for example, colonial plants such as strawberries. They proliferate through subterranean runners, which are tubes of tissue emerging from the plant's underground base. These runners grow sideways, away from the stem, just underneath the soil surface. Unlike roots, these runners also sprout new shoots that grow into aboveground plants. What appear as different plants on the surface may thus effectively be one, consisting of many parts linked through a vast network of runners. What is the organism here? Because aboveground parts and runners originate by cell division, they are not only linked but also genetically nearly identical. Should we call them one whole? Perhaps not, because this whole's "parts" may compete for sunlight, and they certainly can survive and reproduce independently.

Even more familiar are social insects like bees, ants, and termites that live in colonies. Their individuals belong to different castes, for example, workers and soldiers that perform different tasks in the colony. They cooperate in gathering food, raising their offspring, building their nests (often true architectural marvels) and defending them. Only some of the individuals reproduce. All others sacrifice their lives for the greater good. Is this an organism or a colony? If it is an organism, what are its organs? If it is a whole, what are its parts, the individual insects, or the groups we call castes?

Choice Boundaries

Such examples evoke strong opinions about true wholes. One person may call a colony of polyps a whole, but not a colonial plant. Someone else may not be ready to view a slug of amoebae as a whole but will concede that status to a colony of insects. Yet another person only calls organisms like bacteria true wholes.

Consider that the choice is entirely yours. There may be no one truth in this matter.

Does this mean that your choice is completely arbitrary, that it doesn't matter what you choose? By no means. Indeed, this is the most dangerous pitfall of taking the power of choice seriously. To begin with, you probably cannot help making some choice. And what matters in this choice is what you consider important and valuable (a perspective influenced by past choices). And if many people agree with your choice, this choice may carry additional weight.

What does this mean in terms of the organisms I have described, such as the polyps and others? Your choices allow you to ask questions that yield meaningful answers. Why do brainless pumping stomachs and sterile insects help one another? Only viewing the colony as a whole can lead to an answer. Why do polyps fight over food or colonial plants compete for light? Only viewing polyp and plant as the whole will yield an answer. The high art in choosing is to remain aware that past choices are just that, not to cling to one of them, to retain balance on the edge of a sword.

Unnatural Kinds

Wholes have boundaries that separate them from the rest of the world, like the boundaries of time and space separating you and me. For many wholes, these boundaries are not thick, impenetrable brick walls but instead thin, porous membranes. Viewed from the right perspective, they may dissolve entirely. Let us examine these boundaries for wholes like biological species.

Bacteria live in immense communities. A spoonful of garden soil may contain hundreds of kinds of bacterial species.[28] It may contain more individuals than our planet hosts people. It is truly a small universe. This society of bacteria harbors every conceivable lifestyle, ranging from murderous competition to self-sacrificial cooperation; it knows arms races, freeloaders, communal living, cheating, labor sharing, and many other behaviors we unthinkingly consider the human prerogative. You could fill as many libraries about this society as about human society. Here, I want only to highlight one of its facets, a strange gift bacteria possess, the gift of exchanging genes.

Diets distinguish many lifestyles in bacterial societies. Some bacteria thrive on sugars like lactose; others gobble up certain acids; yet others feed on minerals, on the excrement of their fellow bacteria, or on the dead. What makes all these bacteria different? Proteins and, ultimately, genes. To digest food a bacterium needs enzymes, proteins that transform the food into a useful chemical form. Lacking genes for these enzymes, the bacterium would starve at the most sumptuous banquet. Genes and enzymes are also responsible for other bacterial features, including whether a bacterium can live around oxygen (many cannot) and whether a bacterial community can withstand an invader's chemical attack or fend off a viral parasite.

Like other organisms, bacteria pass copies of the genes required for their lifestyle to their offspring. But many bacteria also possess an additional, unusual gift: they can transfer genes between cells, even between cells of unrelated species. They do so either by receiving donations of genes from other cells or by incorporating DNA from dead bacteria. The transferred genes may help the bacterium digest certain foods, fend off chemical attacks, or withstand deadly viruses. To see how truly bizarre this gift is, just imagine that you could obtain genes from other organisms; genes, perhaps, that might allow you to produce energy from sunlight, genes immunizing you against diseases like polio, or genes that would allow you to eat wood.[29]

Gene transfer helps bacteria survive, but it also blurs the boundaries between bacterial species. Let me first say a few words about species. Grouping things into different categories is essential for organizing our world. We do it every second of our waking life, a life that would be unthinkable without such organization. If you look up from your reading, you will see many objects that your mind automatically names and classifies. This is how we make sense of the world. It is a process that begins in the back of our eyes, where our retina creates simple shape categories like circles and bars and ends with complex mental categories like "grand-mother," "banana split," or "elementary particle."

Categories for the mind-boggling diversity of the living are no exception. Thus, while bacterial species may appear different from wholes like bacterial cells, humans, slime molds, or proteins, they are not. All of these are concepts emerging from the incessant yet silent activity of our minds. Everything I talk about here — everything anyone *can* talk about — is built on categories, created by our mind's choices.

Perhaps the only difference between categories like "living cell" and "species" is that grouping organisms into species requires more conscious effort, because the ways of classifying the living are virtually inexhaustible. One could classify organisms by shape, color, or taste; by the number and kind of appendages (legs, scales, flagella); by their ability to lay eggs; and so on. Naturalists labored for centuries to find the best system of organizing life's immense diversity. Eventually, many of them agreed that such a system should reflect an organism's position on the tree of life, its evolutionary history. Such classification not only organizes our knowledge about the living but tells us whether two organisms belong on the same branch of life's tree. It places apes closer to humans than, say, mice, flies, or bacteria, simply because the common ancestor of humans and apes lived more recently than that of humans and mice, flies, or bacteria. Organizing life around this tree, however, poses a challenge. It means reconstructing the history of life itself.

This task may seem hopeless, given that there are millions of species. It would certainly have been impossible in the nineteenth century. But no longer. The key is access to DNA and to genes. To reconstruct history, one can take advantage of the occasional copying errors that dividing cells make when passing on their DNA.

Consider a particular gene — any gene — that is passed on through the generations. In any given generation, the chances of a one-letter change in its DNA text are minute. But that gene may be passed on through hundreds, thousands, and millions of generations. If you observed it on its journey through the generations, you would see more and more changes in its DNA text. The number of these changes is proportional to the time the gene has existed. This means that the number of DNA changes in a gene can be used to measure time. Assume, for example, that copying errors change 1 percent of a gene's DNA letters every million years. If 10 percent of these letters have changed since the gene's origin, then the gene has existed for ten million years.[30] Genes can thus serve as molecular clocks, measuring time on a scale unimaginably large for humans.[31]

With this in mind, let's consider a population of organisms — bacteria or otherwise. Such populations sometimes get subdivided. A storm blows a few fruit flies from North America into Hawaii; a glaciation splits a population with expanding wedges of uninhabitable ice; during a famine, a few animals embark on a suicidal exodus through an inhospitable desert and, against all odds, find lush new pastures. Erupting volcanoes, flooding valleys, and rising mountains are other means of dividing populations. Temporary at first, such separation often turns permanent, lasting thousands or millions of years. By the time another stroke of fortune reunites the separated populations, they may have changed profoundly. In organisms reproducing sexually, the females from one population may no longer be attracted to males from the other, because they no longer recognize scents or mating songs as "sexy"; their differently shaped genitalia may prevent them from mating; when mating, one population's sperm may not penetrate the other's eggs; or their offspring may be malformed.

Mutations that accumulated independently in the divided population account for many of these differences. (Because mutations change DNA text at random, they will modify the same gene differently in two separated populations.) Thus, even when reunited, the two populations will now evolve separately. Their individuals are not of the same kind anymore. Two new biological species have formed.[32]

Any two species alive today, however different, share a common ancestor in some population of the distant past. To reconstruct the tree of life is to find out when this common ancestor lived. But this ancestor is long dead. How to trace it in the rubble of life's several-billion-year-long history? To answer this question, the molecular clocks I mentioned earlier come in handy. Two species with a common ancestor contain many gene pairs — one gene in the first species, the other in the second species — that share a common ancestor. Each member of a pair is a molecular clock that ticks independently of the other member. Simply by comparing the two copies to each other, by counting how many changes their DNA text suffered, one can estimate when their common ancestor lived. This principle is a cornerstone of biology's effort to organize the living. Without it, the origin of many species would be lost in deep time.

Despite being seductively simple, this principle sometime fails, especially for bacteria. You might compare a particular gene in two different bacteria and find that 30 percent of its DNA letters are different in the two species. You would conclude that these species originated many millions of years ago. Yet you might find another shared gene telling a strikingly different story. This gene might be identical in the two species, suggesting that the species have formed recently. One of the genes must be lying: the bacteria's common ancestor could have lived either many millions of years ago or yesterday but not both. The explanation is that one of the genes in the identical pair has recently been transferred between the species.[33]

Exchange of genes between bacteria poses serious problems in reconstructing their history and placing them on the tree of life. If frequent enough, gene exchange may impede reconstructing much of life's past, for earth harbors many more kinds of bacteria than plants and animals combined. Unfortunately, gene transfer is by no means rare. It is more like a habit: bacteria with hundreds, even thousands, of imported genes are the rule rather the exception.[34]

This problem — aside from being a huge practical nuisance — points to a much deeper conundrum. What is the *true* evolutionary history of a bacterium that has imported a thousand genes? Where does it belong on the tree of life? (Is there perhaps no such tree for bacteria?) And how do we delineate the whole, the bacterial species? This problem would not arise if bacteria exchanged only protein parts. It comes only with exchanging genes. Genes are much more stable than other parts, for they preserve their identity — not only in matter but in meaning — over millennia. They also provide bacteria with their distinct lifestyle. We might hope that similar lifestyles reflect similar evolutionary histories. Extensive gene transfer crushes that hope.

This problem exists not only for bacterial species but also for individual bacteria. For if two bacteria have exchanged a thousand genes, does that not blur some boundary between them? The question "what is the whole organism?" then arises again, as it did for colonies of polyps, slime molds, or social insects.

Whether applied to individual bacteria or to bacterial species, the answer is the same: the whole does not exist outside a human choice. Again, this does not mean that choosing is arbitrary or pointless. For example, you might want to find out whether a well's water is contaminated, perhaps by human excrement. Could somebody die from drinking this water? To answer this question, you need to identify the disease-causing organisms. You do not care how many genes they have exchanged as long as you know how to tell them from other bacteria.

Or consider bacteria that feed on heavy metals toxic to us. You might want to use such bacteria to clean up soil that has been contaminated by past mining activity. All you have to do is to identify these bacteria reliably. Such choice may even be useful for reconstructing history. For instance, you might choose to classify bacteria according to the genes they harbor. After all, it is these genes that endow bacteria with their lifestyle and identity. Then you could choose any one gene to reconstruct history. But the history of what? Not the history of the bacterium you see in the microscope, because that one is a mosaic of imported parts. Perhaps all you can reconstruct is the history of the gene and that of the associated lifestyle.[35]

Blurry Species Boundaries

Earlier we encountered the notion of a biological species — organisms that can reproduce sexually with each other. This notion does not apply to bacteria, which are strangers to the kind of sex we practice. Yet for organizing sexually reproducing organisms on earth, the idea of biological species is essential. Unfortunately, it shares all the limitations of other concepts we have encountered.

Imagine a rugged mountain range rising many kilometers above an arid plain. Time has carved into its flanks several deep valleys encircling the range. Perched on these flanks — halfway between valley bottom and a rocky moonscape — lives a species of small, sexually reproducing animals (salamanders, snakes, or squirrels — pick your favorite). Unable to survive extreme cold, they cannot cross over a saddle from one valley to another. And they also cannot cross through the arid and hostile plain surrounding the range. But they are well adapted to living on the mountains flanks, and so they stay. They form several isolated populations, one in each of the valleys encircling the range.

Do these populations belong to the same species? To find out, you bring together individuals from two populations and ask whether they can reproduce with each other.[36] You find that individuals from *any* two populations in adjacent valleys reproduce easily, forming abundant litters that soon have offspring on their own. They clearly belong in the same biological species. To your surprise, however, individuals from distant valleys cannot reproduce with each other. They may have no offspring or their offspring may be malformed or sterile.

Biologists call such populations ring species, because they involve a ring of isolated populations—whether a ring of valleys encircling a mountain range or, perhaps, an archipelago of islands in the sea. Examples include populations of colorful salamanders in the California Sierra Nevada, herring gulls in northern Europe and North America, and greenish warblers on the Tibetan plateau.[37] Individuals from adjacent populations can reproduce, in contrast to individuals from distant populations.

What *is* the species in a ring species? Although any two adjacent populations can reproduce—and thus belong to one species—more distant populations cannot. They belong to different species. The notion of a biological species fails utterly here. No matter how you twist and turn the matter, you cannot organize these organisms by applying a mechanical rule. You have to exercise judgment. You have to make a choice.

The limitations of these two ways of organizing the living—through a common history of genes or through common reproduction—expose the necessity of choice to delineate wholes and to categorize the living. And consider that these limitations, though specific to the living, point to a more general principle: *any whole's boundaries, viewed from the appropriate perspective, disappear.* Consider that this holds for anything I have talked about, anything I could talk about, from the large—humans and their societies—to the small—molecules and atoms.[38] Creating boundaries to organize the world always involves choice, whether we make ourselves aware of it or not.

Once we have made such a choice, many other questions arise about the relationship between a whole and its parts and how this relationship creates the properties of an object or the behavior of an organism. In answering these questions, choice again plays a key role, which I illustrated with only a few examples, such as proteins and swimming bacteria. But the same holds for all conceivable parts and wholes I did not talk about. They include closed canopies of tall trees that choke tiny saplings struggling for light; submicroscopic parasites killing people in the millions; economic depressions devastating families; imported pests wreaking havoc on farmers; technological revolutions germinating in somebody's garage; and submicroscopic genes steering the behavior of animals a billion times their size. All of these wholes and their parts point to the same principle. Parts and wholes are like two sides of the same coin. They are completely separate yet inseparable, for if examined carefully, any part can shape the whole and vice versa.

If you want to understand the relationship among parts and wholes, you have to make choices — just as you must when organizing the world into wholes. That is, if you choose to ask particular questions or to have a specific conversation, you must select a particular perspective. Are these choices arbitrary? Yes and no. Yes, because somebody else may choose to ask a different question. No, because some choices are more useful than others for any one conversation. To identify disease-causing bacteria in drinking water may not require you to ask about genes, protein motors, or whirring flagella. To study bacteria's swimming, lifestyle, or evolutionary history, however, you had better focus on these parts.

In talking about wholes and parts, I focused on conversations that organize the world, scientific conversations and others. Such conversations illustrate how scientists' choices have transformed our world. But the principles I described hold for virtually all human choices. And every choice you make builds on how you have organized your world. What you consider good and bad, right and wrong, how you align yourself politically, the kind of life you want to live, the dreams and hopes you have for you and your family — they all build on how you orient yourself in the world, beginning with your earliest days and even with your ancestors, perhaps as far back as tiny cells propelling themselves toward their fate with whirring submicroscopic engines.

For wholes like folding proteins or swimming bacteria it may be easy to agree with the perspectives I proposed. But these perspectives also apply to you and your role in the world. You can choose to view your life as run by forces beyond your power, your genes, perhaps, or society; or you can choose to take your destiny into your hands. Neither part nor whole completely rules the other, and the space between them is wide open.

CHAPTER 4

Risky Refuges

One million of us, then, die annually. Out of
this million ten or twelve thousand are
stabbed, shot, drowned, hanged, poisoned,
or meet a similarly violent death . . . The
Erie railroad kills 23 to 46 . . . the rest of that
million, amounting in the aggregate to that
appalling figure of 987,631 corpses, die
naturally in their beds! You will excuse me
from taking any more chances on those
beds. The railroads are good enough for me.

Mark Twain, "The Danger of Lying in Bed"

Vast ills have followed a belief in certainty,
whether historical inevitability, grand
diplomatic designs, or extreme views on
economic policy.

Kenneth Arrow

Is your life risky? Are you a war correspondent, a test pilot, a circus acrobat, or an undercover police officer? Alternatively, are you an accountant, an administrator, or a receptionist? (The ceiling could perhaps cave in on you. But what are the chances of that?) How do you feel about the world changing frantically all around you? Are you politically conservative? Do you want to preserve and enshrine your ancestors' way of life? Or would you rather dump their dusty traditions? Do you feel that younger people resemble space aliens rather than humans? Or do you embrace their boldness and accept every cultural, technological, or political innovation you encounter?

To take risks is to allow change. To preserve the status quo is playing it safe. Or so it seems. Can you live without taking risks? Is it sometimes risky to play it safe? Is it sometimes safer to take a risk? These are the opposites that will occupy us next: risk and safety, change and stability. They harbor the same tensions we have encountered previously.

As usual, we are not the first to uncover these tensions. Before us came others. Billions of years before us. After exploring what these others have learned, we will turn to the perspective these tensions give rise to. It is a perspective on the adventure that all of life is: whatever risks this adventure requires, in the long run it creates freedom, the freedom from earlier constraints, the freedom to choose a way of living and to create a world that allows this way of living.

Necessity, the Mother of Russian Roulette

To study bacteria, you can grow them in shallow petri dishes filled with a food-laden solid jelly. An individual bacterium cannot swim on this solid and will thus stay put. But it divides, and so do its offspring and their offspring, until its descendants have formed a colony visible to the naked eye. The size of a pimple, this colony can contain hundreds of millions of bacteria. And a single petri dish no larger than your palm can contain hundreds of such colonies. Each harbors the descendants of one founding bacterium.

The bacteria in each colony have voracious appetites and divide so rapidly that a colony takes a mere two days to form. Its millions of inhabitants swiftly consume the food around them. Days after having formed, the colony begins to starve. Fortunately, bacteria have developed sophisticated strategies to evade starvation. One is a peculiar form of Russian roulette: some starving bacteria gamble with the ultimate risk, death.

Each of the bacteria in a colony harbors a long string of DNA containing thousands of genes. This string originated, copy after copy after copy, from the pioneering cell that founded the colony. The copies, however, are not perfect. If you compared genes across the colony's members, you would see an occasional difference in their DNA. Where most bacteria might contain a G, some might harbor a T; where most contain an A, some may harbor a C; and so on. These mutations occur at random positions in the DNA.[1]

A colony may harbor different variants of any one gene. How many? That depends on the colony's food supply and thus on the colony's age. A young, well-fed colony of one hundred million bacteria might contain fifty variants of each gene.[2] An old colony of the same size, however, might contain thousands of variants. Old colonies can show an explosion of DNA variation, even though many of their cells do not divide. They do not copy their DNA but only cling to life. Their DNA changes so drastically because of changes in a particular class of proteins known as DNA repair proteins.[3]

In all cells, starving or well fed, DNA is constantly exposed to external influences that erode its molecular code. Some of these influences penetrate the cell from the outside: particles from high-energy radiation, such as ultraviolet light from the sun, smash into the cell's atoms and change their structure. Other eroding influences originate from inside, such as the waste that cells incessantly spew. In producing waste, cells are similar to human factories and power plants. And like the excretions of human factories, some of a cell's waste poses fatal hazards. It consists of enormously reactive molecules called free radicals, that can destroy about any other molecule (and themselves with it). To a free radical, a DNA letter is just another molecule. The radical will react with this molecule, change it, and thus alter the information in a gene.

Unchecked, both external influences and internal waste would slowly alter any DNA molecule and eventually reduce it to incomprehensible gibberish. Altered DNA translates into altered proteins that contain different amino acids and fold into different shapes. Such shape changes would disable many proteins and have fatal consequences. For instance, many proteins are involved in essential and ongoing cell maintenance, which is required even for starving cells. They protect the cell's delicate membranes, other proteins, and the cytoskeleton from the very hazards that damage DNA. To disable these proteins is to kill the cell.

Cells must stem this destructive tide of change. They do so with the aid of DNA repair proteins. Every living cell harbors such proteins; their task is to detect and fix DNA errors. In a growing, well-fed colony of bacteria, repair proteins fix the vast majority of DNA changes. Cells in a starving colony, however, alter this repair crew. They produce sloppier repair proteins that introduce new errors while fixing others.[4] As a consequence, more and more DNA changes accumulate in these cells. This is how DNA change accelerates in starving cells.

As unlikely as it seems, there may be method behind such sloppy DNA repair. It may be a gamble with enormous stakes.[5]

Cells need enzymes to transform food molecules into energy and building materials, and starvation renders these enzymes useless. But some plentiful molecule around the colony might serve as food if only the cell's enzymes could recognize it and transform it into a useful molecule. This molecule would solve the cell's problems. Yet as long as a cell cannot "see" it as food, this molecule might as well not exist. Some enzymes' surfaces need to change in just the right way to recognize this molecule's surface.

By changing a cell's DNA at random, the cell changes its protein content, gambling with its proteins. Most of the changes will disable enzymes. But a tiny fraction might lead to a molecular Eureka! experience, the creation of an enzyme with the ability to digest novel food.[6] A cell changed in this way has won the ability to grow and divide, whereas all others starve to death. It becomes the founder of a new colony of millions.[7]

Consider this analogy. A population of a billion people in some hypothetical developing country faces a fuel crisis. Their cars need gasoline, but none is available. Perhaps these people are a little like me, not knowing the first thing about cars. Desperate, they all start to modify their engines haphazardly. They take parts out, modify others, and add new ones, all in the hope that these arbitrary changes would allow the cars to run on more abundant fuels such as vegetable oil or alcohol.

Their chances of ruining the engine are very high. Their chances of success are extremely small.[8] For the unplanned creation of new combustion engines, we do not know how small. But for bacterial colonies, the odds are estimable, because we can easily study a billion bacteria. We can pose a question by giving a colony unknown food and then observe how many of its bacteria learn to thrive. The members of a well-fed bacterial colony with a hundred million cells would take many years to do so. But in a gambling, old colony, this process can be accelerated many thousand-fold, taking less than a day before the first cells recognize the new food.[9]

A starving cell's gamble speaks to other tensions we have encountered. For instance, a cell ("self") most likely does not benefit from shutting down DNA repair. Most likely, it will die. In contrast, the entire colony ("other") benefits immensely when one cell survives and becomes the seed of new life. Cell and colony are thus intimately linked in fate.

The Safety of Risk

For now, just consider this question: What is risky and what is safe — from a cell's perspective? To maintain DNA repair is certainly safe. Cells that keep up repair certainly will not die from corroding cell parts. To maintain DNA repair, however, may also be very risky: the cell's inaction tacitly assumes that food will again become available. The chances of that happening may be nil. Conversely, shutting down DNA repair is obviously risky. It will eventually destroy the cell. But is it perhaps the only safe course of action? It might create the one and only survival strategy, the end of starvation and the formation of new life.[10]

In hindsight, the safe approach that has helped bacteria survive for more than three billion years is to gamble when facing death. This is the strategy bacteria pursue today, even though it ensures that all but a few cells in a starving colony will die.[11] Hindsight, however, does not give us the one *true* answer to the question what is safe or risky? After all, from today on, the bacterium's time-honored strategy may no longer work.

Consider that the one true answer does not exist — that risk and safety, despite being opposites, are as inseparable as two sides of a coin. Anything that is at first glance safe involves incalculable risks. Anything risky, when viewed from the proper angle, may provide a safe haven.

But isn't there one correct angle, one ultimate and true perspective? Consider that the answer is no. Consider that the choice of perspective is entirely up to the decision maker, whether it is a bacterium or you. Risk and safety contain a paradoxical tension similar to those we have encountered earlier. Only a choice—however inadequate it may turn out—can resolve this tension.

I will first address the most obvious objections to this position. Pretend that we know enough about the bacterium's world—what foods it harbors, how often new food sources arise, what enzymes the bacterium produces, how these enzymes recognize food molecules—to calculate that bacterium's risk of dying when it gambles. This calculation would be no different—though more complicated—from calculating the odds of losing in a game of cards, dice, or coin flips. (It is riskier to wager on getting "heads" three times in a row than getting "heads" once.) Having complete knowledge of the circumstances determines risk. We can thus truly evaluate risk, at least *in principle*. To determine risk does not require choice.

What is wrong with this argument? First, and least problematically, risk is a matter of values.[12] The potential gain or loss in a risky decision must be of value to the decision maker. Second, to take an outsider's point of view is to cheat. The problem must be viewed from the perspective of the decision maker, be it a human or a bacterium, and its power to evaluate risk. And third, having complete information—knowing all that might matter for a decision—is impossible even in principle, for either the bacterium or a human.

To approach the problem of incomplete information, let's explore human decision making. The topic is vast and affects all walks of life: how to determine harvest dates using weather forecasts; insurance premiums for a priceless painting; a safe place to build a home; where to drill for oil; the purchase price of common stock; the risk of war; and so on. These and countless other problems fill entire libraries of calculations in decision theory, risk analysis, insurance mathematics, and game theory.[13]

Fortunately, only a few points are germane. First, it is crucial to distinguish between idealized games of chance like flipping coins and decisions in life. Some students of risk distinguish the two via the notions of risk and uncertainty.[14] In a decision involving risk you know the odds, such as when betting on a coin flip or when playing roulette. In this sense, you have complete information about the problem at hand. In a decision involving uncertainty you do not know the odds. Examples include the prediction of earthquakes, outcomes of elections, or stock market movements.

Risks are calculable only where you know the odds. All such calculations are identical in spirit—though perhaps more complicated—to calculating the odds in a game of chance. But virtually no real-life decisions are like that. Real-life decisions cannot be reduced to games of chance with calculable risks. We often pretend to know the odds, but numerous factors ranging from the weather to the behavior of other people prevent us from doing so. To use the distinction between risk and uncertainty, all real-life decisions involve uncertainty. (For my purposes, however, I will continue to use "risk" as the all-encompassing word.)

It is perhaps no coincidence that it was economists, such as John Maynard Keynes or Nobel Laureate Kenneth Arrow, who first recognized the abyss between calculable risks and real-life decisions: economic decisions are important in everybody's life, and also involve much uncertainty, yet have rather simple consequences, such as to win or to lose money. Many other decisions we will encounter have ramifications vastly more complex.

Calculable games of chance cannot guide real-life decisions, although we can fool ourselves into thinking that they do. Nevertheless, we like to use some information — however incomplete — to make decisions. An anecdote from economist Kenneth Arrow captures the essence of this dilemma. During World War II, Arrow forecast weather for the U.S. Air Force. His team found that their long-term weather forecasts were no better than random numbers pulled from a hat. They thus asked their superiors to be relieved of the task. Their superiors replied, "The commanding general is well aware that the forecasts are no good. However, he needs them for planning purposes."[15]

Another general point about risky decisions concerns the information we use to guide them: it is all information about the past. Take the decision to buy stock and the associated risk of financial loss. Some people buy stock on a hunch. Most others — especially professional investors — use information about the stock's past to predict future risk. It could be the history of a stock's performance, the ratio of stock price to past company earnings, or stock volatility, the extent of a stock's past price fluctuations. These and many other risk indicators evaluate future risks based on the past. All fail frequently, as reflected in the stockbroker's mantra that past performance does not guarantee future results. In addition, their failure is not limited to major stock market crashes. Take the case of actively managed stock funds, in which a professional manager is paid to assess the risk of holding a particular company's stock. The majority of these funds yield returns worse than the stock market average, although humans with expert knowledge manage them.[16] The average decision based on past stock performance is no better than a random number pulled from a hat.

That past experience guides future decisions holds not only for financial markets but everywhere in life. (It holds even for games of chance, because only past experience tells you that a coin flip yields "heads" half the time.) And the past—even though it is all we have—has proven to be a poor indicator of the future.

In many human games, the consequences of decisions are clear. This holds even for games as complicated as the stock market, where you can either win or lose money.[17] In contrast, in many games played in nature, decisions can have unanticipated outcomes, such as the ability to digest a new food. The outcomes of such games can change the game itself, and they may even create new games. Consider the genetic gamble that organisms have pursued for billions of years. The outcome has been more than a million species that face different risks and thus play different games.

For all these reasons, I argue for the perspective that safety and risk, though opposites from one angle, are inextricably linked like the two sides of a coin. The price you pay for making a seemingly safe decision is that this safety itself harbors unanticipated risks. Conversely, a risky decision may buy you the chance to escape an otherwise fatal situation. Consequently, decisions on what is risky and what is safe inevitably contain elements of choice, because who can be the judge, except the bacterium or the human decision maker? And if neither can calculate risk, not even in principle, who can?[18]

We and the gambling bacteria have a lot in common, for neither of us can foresee the outcome of our decisions. In fact, our own myopia makes gambling bacteria look good by comparison. A living bacterium is like a stockbroker who has made only good decisions for the past three billion years. It did not make a single wrong call, for any wrong call would have meant its death. But like a stockbroker, its next call may fail. What worked in the past ("when desperately starving, shut down DNA repair") may work no longer.[19]

Safety through Sloppiness

Starving and gambling bacteria are not extreme or unusual. The same principles apply throughout biology. For example, consider an intriguing difference in the abundance of mutations among species.

Recall that DNA can change for a variety of reasons, including unfaithful DNA copying, external radiation, and internal waste. Cells can fix DNA changes using repair proteins and thus control mutations to some extent. Curiously, the mutation frequency—the number of mutations that occur every generation—varies enormously among species.[20] Some organisms accumulate DNA changes at a thousandfold lower frequency than others. Even organisms relatively similar to each other, such as different species of bacteria, show manyfold differences in mutation rates. As a rule of thumb, the more DNA and the more genes a microbe has, the lower the number of mutations each gene accumulates per generation.[21]

A possible explanation for this pattern is that organisms with more genes also have more proteins, and each broken protein may mean death to the organism. Because most DNA changes will break proteins, organisms with more proteins may thus tolerate fewer mutations.[22] However, the actual explanation for varying mutation frequencies is secondary to our conversation. More important is the question it raises about risk: If cells can control DNA repair, why are mutations not equally rare in all organisms?

At first sight, allowing as few mutations as possible would seem the safe thing to do. Any organism alive today embodies a successful strategy of making a living. The lifestyle of bacteria has been successful almost as long as life itself—more than three billion years. Why change a winning team of proteins? Why allow mutations that may disable proteins?

First, organisms cannot fix all DNA changes. Fixing genes takes time and energy. Fixing too many genes costs too much time and energy that are needed for other vital activities, such as swimming for food. But some organisms clearly fix more mutations than others. Cost thus cannot be the whole answer.[23]

Here is a second possibility. There is no guarantee that an organism's lifestyle will be successful forever. The world around the organism is constantly changing and has been doing so for more than three billion years. An organism that allows rare changes in DNA, on the off chance of inventing an improved, more successful lifestyle, always gambles, if only a little bit at a time.[24] What starving colonies of bacteria display in the extreme may thus hold for many other organisms. Just as well-off investors could be even better off, so could organisms with a successful lifestyle be even more successful. The better is the enemy of the good. Embodying this

principle, the living may thus accept a gamble's risk to reap its rare benefits. Starving bacteria, whose lifestyle has turned unsuccessful, may simply
gamble more intensely than other organisms, on the off chance that the
gamble will endow them with a life-saving innovation.[25]

An Ever-Branching Tree of Lifestyles

We have explored that mutations change lifestyles. But how do lifestyles
themselves change? And what are the opportunities and risks that come
with such change? Before we examine these questions, consider that any
one task in life can be solved in innumerable ways and that many successful solutions build on the success of other solutions.

The task of acquiring energy provides a vivid example. The more energy an organism can access, the faster it can grow and reproduce. Plants,
for instance, employ innumerable strategies to acquire energy. Most of
these strategies rely heavily on light. The more light a plant can access, the
better it will be able to grow and reproduce. To access as much light
as possible, many plants thus grow into the sky. But doing so requires
support against gravity and wind. To provide this support, many plants
build a mighty trunk, which requires much energy and building materials.
Couldn't some of this energy be saved? Couldn't one economize and still
access much sunlight? The answer is yes, if one relies on the support of
others, such as other plants' mighty trunks. Plants that pursue this alternative strategy include vines, which are flimsy but strong enough to embrace
their support. Building a vine does not cost much energy, and it allows
rapid growth and access to light.

Yet another, altogether different strategy to acquire energy is pursued
by parasitic plants. The roots of these plants invade the transportation
network of other plants and tap it for energy. This strategy often bypasses
the need for light energy altogether.

Another case in point is the task of obtaining essential minerals like
nitrogen. Most plants rely on nitrogen bound in soil minerals. The problem is that many soils are poor in nitrogen. Is there a way out of this
bottleneck? Yes, because the air around plants consists of some 70 percent
nitrogen. Most plants cannot extract nitrogen from the air. But legumes —
peas, beans, and their relatives — have found access to airborne nitrogen.
They team up with *Rhizobia,* bacteria that can mine the air for nitrogen.

Rhizobia live in the roots of legumes, where they obtain nutrients and transform airborne nitrogen into a form useful for the plants. A third strategy to obtain nitrogen is pursued by carnivorous plants. They kill animals, another rich nitrogen source. Examined further, each of these strategies to obtain nitrogen can be pursued in many ways. Carnivorous plants, for instance, may drown their hapless victims in a pool of digestive juices, sedate them with powerful narcotics, trap them with potent glues, or imprison them in snapping jaws.

What distinguishes the millions of species on earth is how they make a living, how they obtain energy, how they escape predators, how they chase prey, how they survive bad times through dormancy or migration, how they protect their young, and so on. Any of these objectives can be accomplished in innumerable ways, nested within one another, that breed even more strategies of making a living. Still more ways of making a living build on these strategies. We can enter this hierarchy at any one point and explore how lifestyles change with a few examples.

Dangerous Indifference

Driven by mutations, organisms and their lifestyles change slowly over time. Such changes bear obvious risks to the individual: many harm the organism, such as the changes that stall a bacterial engine or stunt a virulent parasite. Other changes — gaining the ability to digest new food or kill prey — are obviously beneficial. Yet other, *neutral* changes apparently do not affect the organism at all.[26] Organisms probably have little or no foresight about the effects of mutation on them.[27] If, however, they had such foresight, they would strive to prevent deleterious changes, be indifferent to neutral changes, and favor beneficial changes, or so we might think.

But first appearances are, as so often, deceptive: if an organism had *sufficient* foresight, it would despair at evaluating whether a change is for the better or for the worse. The first of several cautionary tales is about neutral changes, changes that do not affect an organism when they first occur. Such changes need not stay neutral forever. Long after they occurred, perhaps millions of years later, some neutral changes can abruptly turn fatal.[28]

Bacterial cells and human cells need very similar molecules — some of them proteins, others small food molecules—to help them grow and divide. Bacteria can build virtually all of these molecules from tiny building blocks like minerals and sugars. Our cells, however, cannot; the production of vitamins, complicated but essential molecules, eludes them. Our cells thus obtain more than a dozen "vitamins" that bacteria can produce on their own through the food we eat. Bacteria have the necessary enzymes to produce these molecules, and we do not. But why?

Long ago in life's history, bacteria and we shared a single-celled common ancestor. Like today's bacteria, this common ancestor could produce vitamins. However, sometime since human and bacterial history parted, our ancestors lost this ability. Mutations corroded the necessary genes. We could get away with losing them because our food supply always contained these vitamins. The mutations went unnoticed because our ancestors did not rely on the disabled genes. They were mutations that rendered the loss of vitamin production a neutral lifestyle change.[29]

This simple principle—get rid of whatever you don't need—will be painfully obvious to anyone whose garage bursts at the seams with treasures from times gone by: magazines, tools, toys, you name it. (Everything but the car, which no longer fits.) Periodically one purges those treasures that have turned to trash. Genes suffer the same purges on an ongoing basis. Dispensable genes, genes that have become superfluous, are purged through mutations that disable and eventually destroy them.

Vitamin production is by no means the only lifestyle feature subject to purging. The most striking cases of purging occur in parasites, especially those living inside other cells, where hosts may serve them not only a handful of vitamins but almost all of life's necessities. As a result, some parasites have become incapable of producing the building blocks of their own proteins, others of producing the building blocks of their own DNA; some have lost the ability to build their cell wall, others the ability to respire—to make energy using oxygen, and others still the ability to make energy altogether; and some parasites have lost many of these abilities and thousands of genes with them.[30] Despite all these changes, they live happily ever after. For as long as the host provides, these changes are neutral.

In addition to material services, many animal hosts provide their parasites with the ability to see, to hear—through the host's sensory organs—and to go places. Such parasites can thus safely get rid of these abilities. The minimalists among them keep only the barest of necessities, the ability to reproduce. Their bodies become reduced to sacks of sperm- or egg-producing tissue, nourished by the host through a filigree of feeding tubes.

In parasites that cause human disease, a particular kind of purging occurs. Bacteria such as *Mycobacterium tuberculosis,* which causes tuberculosis, have become resistant to multiple drugs.[31] Any such resistance gets lost rapidly when the respective genes have become dispensable, when doctors no longer apply the drug. Losing resistance may then become a neutral lifestyle change.

Not only parasites purge antiquated lifestyle features. Free-living organisms do as well, such as those that dwell in the everlasting darkness of subterranean caves. They do not need to see, so they lose their eyes. Fish moving onto land lose their gills. Some burrowing organisms lose their legs, others their tails once they are not needed to provide equilibrium. And so on. In sum, everything from individual proteins to complex organs, once inessential, is subject to swift purges. All such purges can be neutral.[32]

What does this have to do with risk and safety? Nothing, if the world stays the same; everything, if the world changes. And the world changes all the time. Which of the stuff in the garage can you safely throw out? Hard to tell, isn't it? Perhaps one day, that old skateboard might come in handy if your car is in the shop. And what about the bowling ball that has been gathering dust for years? That thing was expensive! Who knows whether the kids might take up bowling one day?

Similarly, no organism can know whether a purged part may become useful some day. Except that the price for throwing out the wrong item is usually death. *Mycobacterium tuberculosis,* for instance, may have purged its ability to disarm a drug, a neutral change in the drug's absence. But many generations down the line, doctors may decide to use that drug again. And what about the parasite that depends fully on its host, whether for orientation, movement, or food? What if the host learns to defend itself, either locking the parasite out or evicting its tenant through its immune system? Imagine the situation of a parasite that lost everything but its genitals. Once evicted, what is it to do?

To come back to our opening example, vitamin deficiencies can also spell death. Vitamin A deficiencies, for instance, are a huge public health problem in regions where rice (lacking vitamin A) has become a dietary staple. There, vitamin A deficiencies can lead to blindness and death, and nearly a billion people are affected by such deficiencies. Purging the ability to produce vitamin A may have been neutral in our ancestors' history millions of years ago.[33] But it is far from neutral for a person dying from vitamin A deficiency today.

The Innovative Power of Indifference

These examples show that any one of the world's many capricious turns can render neutral change disastrous. But the converse also applies. Originally neutral changes can yield invaluable benefits. Take as an example how insects — bees, dragonflies, wasps, and others — learned to fly.

We know the answer to this question for birds and bats. Their ancestors walked on four limbs, two of which — the forelimbs — slowly changed shape. Their forelimbs evolved a surface sufficiently large to glide and eventually to fly. But insect wings are very different from bird or bat wings: they are not transformed limbs, for both flightless and flying insects have six legs, which means that no legs have been transformed into wings. In addition, mammalian forelimbs are very similar to bird and bat wings in their bone organization, whereas insect limbs bear no semblance to wings. So where did insect wings come from?

Some think that insect wings started as useless little flaps of cuticle — an insect's version of skin — on a fly's back. Now, organisms do not generally evolve features as complicated as wings overnight. However, minor changes like more or less hair, supernumerary or fused fingers, or extra flaps of skin are common. They are tolerated as long as they are neutral. (Being neutral, they would also eventually disappear again, unless unanticipated uses arise.) Others argue that wings originated as leg appendages that helped the water-living ancestors of flying insects — relatives of lobsters and shrimp — breathe. Whatever the original use of these appendages, they came about for reasons unrelated to flying. They were neutral changes with respect to the flying lifestyle.[34]

How did such changes permit the leap to flying? Through a series of small steps, many of them also neutral with respect to flying. Although we may never know for sure, the first of these steps may have looked like this. Featherweight insects can live on a water surface without sinking; they can literally walk on water. But walking on water uses energy they could expend on other activities. We already saw — in the case of fast-growing yet flimsy vines — that one way of having more energy is to spend less. The same applies to insects. On windy days, a water-living insect can save energy by sailing with its cuticle flaps — you might call them wingflaps. (On a windless day, of course, it has to walk on water like everybody else.) Once retained, wingflaps might begin to change further. Their surface might increase, allowing faster sailing. Muscles attached to them might allow them to turn with the wind. Generations later, these muscles become sufficiently strong to swing the wingflaps forth and back, thus propelling the insect. Eventually, wind and muscles together might allow takeoff.

A series of changes, individually neutral with respect to the ability to fly, can lead to one of the greatest innovations ever. Of course, any one of these changes, an improvement in hindsight, might have turned out to be a complete disaster: oversized wingflaps might help blow the insect out of the water and into a hostile habitat, or by glittering in the sun, they might advertise the insect to predators.[35]

The Perilous Depths of Time

Not only neutral changes can take a different turn. Beneficial changes can, too. Lifestyles that have been following a steady and successful course can suddenly veer off a cliff. In fact, *almost all successful lifestyles fail disastrously, given enough time.* Unfortunately, to detect good changes that turned bad is difficult, simply because most of those who harbor these changes are dead. They have disappeared from our planet of survivors. Only their remains let us glimpse the terrifying magnitude of this truth.[36]

More than a billion years ago, life on earth underwent a profound transformation: Many-celled organisms emerged from a selfish soup of dividing cells. They were a strange lot, these early many-celled organisms. That, at least, is the story told by a rich fossil fauna from some six hundred million years ago, when fossils first became abundant. These fossils include some familiar faces — organisms that have survived to this day. But among them are also many otherworldly creatures, alien to life as we know it. They seem to have sprung from the overflowing imagination of a science fiction writer, blending features of flatworms, crustaceans, insects, and fish, as if somebody had created new life-forms by throwing all imaginable forms in a box and shaking hard.

Some of the results are truly bizarre wonders, an alien creature with five mushroom-shaped eyes, a proboscis resembling a vacuum hose, and fifteen paddle-shaped gills attached to its sides; an eccentric creation with a lightbulb-shaped head in front and a stubby proboscis at the rear — although which end is the rear is uncertain — suspended on fourteen jointless stilts and wriggling seven tentacles on its back;[37] a tulip-shaped beast clinging to the ground through a structure faintly resembling a toilet plunger, its mouth next to the anus at the flower's bottom. They and many other fantastic creations populated earth for many million years — much longer than humans have so far. Their representatives occurred all over the globe. Thus, by any standard, their lifestyle was extraordinarily successful. But eventually they disappeared, each and every one of them, blown away by the gales of history.[38]

Earth is no stranger to its children's death. Its history has seen five periods of great dying — mass extinctions, each of which killed more than half of life. You may be familiar with some of its victims. Take trilobites, whose remnants litter not only the world's souvenir shops but also the ocean floors. Some three hundred million years ago, the earth's oceans were teeming with these relatives of lobsters, shrimps, and insects. Trilobites were perhaps the most successful water-living creatures ever. But they also became victims — every single one — of the greatest dying ever, a mass extinction occurring some 225 million years ago. In this global disaster, multicellular life came as close as it ever has to being extinguished: 95 percent of marine life and 70 percent of land-bound life disappeared.

The most familiar mass extinction claimed a children's favorite: the dinosaurs. We cannot be certain what was to blame. Perhaps it was the giant asteroid that careened into the Yucatán peninsula sixty-five million years ago, triggering an apocalyptic climate change; perhaps it was a more gradually changing world, including the ascent of mammals. But we do know that dinosaurs were not alone in their dying. Some 60 percent of species — millions — vanished with them.

This mass extinction has a feature also seen in other extinctions. It was selective. Some life-forms — dinosaurs, mollusks, and most land plants — were devastated. Others — ferns, mammals, reptiles other than dinosaurs — remained unscathed or even thrived.

Although mass extinctions were life's true blockbusters, they unfolded against a backdrop of countless ongoing extinctions. These background extinctions began with life itself, and they have continued through the eons. The sum total of their victims may well be greater than those of mass extinctions.

The story of life as told by the past is that the vast majority, more than 99.9 percent of species, will perish. (Past success is truly no indicator of future success; rather, it seems to indicate future failure.) Each of these perished species originated from some ancestor through a change that created success, a successful lifestyle. But whatever the reasons — climate, enemies, meteors — this successful lifestyle eventually failed. The disastrous failures of successful lifestyles highlight how inseparably close are risk and safety, failure and success, despite being diametrically opposed.

Not only fossils convey this message. Recently extinguished species do, too. Take the victims of invasive species, introduced by humans to new habitats, either unknowingly — as stowaways — or deliberately to decimate pests. One case in point is the brown tree snake, a native of Australia and Indonesia. It hitchhiked to Guam on military aircraft in the middle of the twentieth century. This sly predator has nearly exterminated many of Guam's native forest birds, rendering their successful lifestyle, delicately adapted to Guam's rain forests, obsolete in a matter of years.

Another example is the Nile perch. Humans introduced it to Lake Victoria around the same time as the brown tree snake made it to Guam, except that they introduced the perch deliberately to replenish an exhausted stock of native fish. But its introduction backfired. A competitor and predator of native fish, the Nile perch has contributed to the extinction of more than two hundred native fish species.

The water hyacinth, a native of South America, is one of the fastest growing water plants, doubling its population sizes in as little as twelve days. In the United States, it used to be a popular pond plant because of its beautiful flowers. Now it has become one of the world's worst aquatic weeds. Not only does the dense canopy of its leaves — up to two hundred tons per acre — suffocate most other plants, but they can also clog rivers and lakes, bringing shipping to a standstill.

A related terrestrial menace is the ornamental, purple-leaved tree miconia, introduced to a botanical garden in Tahiti in 1937. Miconia was spread into the wild by fruit-eating birds, and its oppressive canopy now covers vast swaths of Tahiti and threatens to exterminate native trees. The Indian mongoose, a native of Iran, was introduced to Mauritius, Fiji, and Hawaii in the late 1800s to control rats. This fast-moving mammalian predator has extinguished many native birds, reptiles, and amphibians.[39]

The list goes on and on. Thousands of invasive species are spreading with humans, and they obliterate innumerable species and their lifestyles. The perished lifestyles illustrate how fleeting the beneficial changes were that created them. Invasive species, together with habitat destruction, are giving rise to what some call the sixth mass extinction. You can think of this sixth extinction as another metaphor for the effects of an ever-changing world. If these effects are unpredictable even to us — who can contemplate them beforehand — how more unpredictable would they be for other organisms?

The ghastly story told by millions of extinct species, including our dead ancestors, conveys the tension between risk and safety in changing lifestyles. But so do the lifestyles of many organisms still alive today. For many of these lifestyles have self-defeating facets, such as that of "effective" parasites that decimate their host. This lifestyle's features often arises through a change — parasites switching hosts — with adverse side effects. However, consider that the same holds for any lifestyle. There is no lifestyle without vulnerabilities, without an opening that may lead to its eventual demise.

The Ultimate Conservatives

A changing world can swiftly turn the greatest innovation into disaster. And our planet changes incessantly. Its continents drift restlessly across its surface. Arctic deserts become tropical swamps, and tropical swamps become arctic deserts. On top of the wandering continents, the climate perpetually shifts, alternating balmy periods with ice ages. New organisms constantly enter this backdrop through successful genetic gambles. In the face of such turmoil, the living's only viable strategy may seem to keep pace. Incessant change — however risky — may seem to be the only refuge. But as usual, appearances are deceptive when it comes to risk and safety. Like ships navigating in dense fog, most organisms keep changing toward an invisible future. But a few organisms drop anchor and, against all odds, weather time.

The ginkgo may be earth's oldest seed plant alive. This tree originated in its present form at least 270 million years ago. Easily identified by the archaic beauty of its leaf veins, the ginkgo changed little through several mass extinctions, including those that swept away trilobites and dinosaurs. Westerners thought it extinct until 1691, when the German Engelbert Kaempfer rediscovered it alive and well in Japan.

Organisms like the ginkgo, of ancient origin and unchanged over many million years, are often called living fossils. They have few or no close relatives and often live only in a small region of the globe. There are not too many of them. They include Latimeria, a lobe-finned fish closely related to the earliest vertebrates conquering land. Thought extinct like the ginkgo, this inhabitant of deep sea trenches was rediscovered in 1938. Sea lilies or crinoids, also inhabitants of the deep, were rediscovered in 1890. The lizardlike reptile Sphenodon, the primitive mollusk Neopilina, the horseshoe crab, and the pink-flowered magnolia are also among this select group of holdouts.

Do these organisms have a secret for success? The answer seems to be no. Different living fossils have weathered time for diverse reasons. Some, like the horseshoe crab, are jacks-of-all-trades; it walks, swims, and burrows well, although not as well as specialists in walking, swimming, or burrowing. Its generalist lifestyle may enable it to cope better with the demands of a changing world. Other living fossils have taken refuge in harsh but stable environments like the deep sea. Still others, like Sphenodon, may simply have won the lottery of an agreeable environment with few competitors.[40] Whatever their reasons for enduring, living fossils show that the obvious solution to surviving in a changing world — change — is not the only one. Their small numbers also highlight the treacherous safety of conservativism.

The Very Long View

Gambling bacteria, stockbrokers, the fleetingness of innovative lifestyles, and the history of life as told by fossils — dead or alive — all make the same case. Being opposites, risk and safety are as different as can be. Yet one cannot exist without the other, and wherever you find one, the other lurks nearby — you just have to look carefully.

Certain knowledge in matters of risk and safety does not exist. Its absence is not just a limitation of our technology or our minds. The question whether any one life decision is safe or risky requires a choice whose merits will be obvious only in hindsight. There is no one true answer.

If you thought big, really big, however, you might counter this train of thought with the following perspective: What if history has an overall direction—toward the better? However many battles our ancestors lost, our descendants will win the war in the end. Eventually, the world will be more hospitable and benevolent, a better place. All our uncertainties are only teething problems, because ultimately the world will become Home.

This perspective is alluring because it soothes our anxieties. The possibility that history is goal-oriented has a name: cosmic teleology.[41] Life and evolution are cosmically teleological if they are oriented toward a goal—however distant, however enigmatic to us. If evolution moves toward the better, then the world is ultimately a safe place. Blunders may happen, often, but eventually things will turn out well. (Although the way things have been going until now, we humans may all be dead by then.)[42]

Although nobody could disprove that history has a direction, the case for cosmic teleology is weak and getting weaker. To see why, let's focus on a specific example, the giraffe's neck. It illustrates some of the key problems with cosmic teleology. The giraffe's ancestors, far from today's graceful swans of the savannah, resembled large deer. In some population of these ancestors, a few individuals had longer necks, just as some people are taller than others. They were able to reach higher into trees, reach more leaves, make a better living, and thus raise more offspring. Long necks being heritable, their offspring also tended to have long necks. The next generation thus harbored more long-necked ancestors. Over countless generations, and over the unimaginable long time of fifteen million years, these slight changes stacked up to the long necks of today's giraffe.[43]

A cosmic teleologian would argue that the giraffe's history *had* to unfold this way. Its path was the one and only path history could take, because it led to the best possible lifestyle. What's wrong with this idea? First, a bit of imagination can offer equally successful alternatives. Why did the giraffe's ancestors not acquire tree-climbing skills to make a living? Climbing seems a more economical way to feed on tall trees, and many tree-climbing organisms live exceedingly well. Or what about taking the route of elephants? Their sheer size allows them to tear off large branches with little effort, even to knock over a whole tree in a pinch. Unlike the giraffe, elephants are not limited by their height. Or why did the giraffe's ancestors not evolve a parasitic lifestyle? After all, parasitism is a spectacular success, shared by the majority of organisms.

To acquire long necks is just one possible path toward a successful living. Moreover, if history is any guide, long necks will eventually become unsuccessful—whether because of some voracious tree-climbing food competitor or another calamity, who is to say? But in addition, a giraffe's lifestyle is far from problem-free. Giraffes cannot sustain high running speeds for long and must thus defend themselves against attackers via powerful kicks front and back, sometimes inadvertently killing their own young by doing so.[44] And have you ever watched a giraffe drinking? It is a truly ungraceful sight, because the giraffe must awkwardly spread its stilted legs to lower its head to the ground. This posture is also fraught with danger from stealthily approaching predators. On top of that, the blood rushing through the neck's long column pressurizes the head dangerously. Giraffe had to acquire sophisticated mechanisms to prevent a fatal buildup of blood pressure.

This example goes to show that success stories are inevitable only in hindsight. Nevertheless, some facets of life's course are predictable. Whales originated from hoofed land-living mammals similar to cows. Whales and fish differ in many respects: fish breathe through gills, and whales have lungs; fish do not lactate, but whales nurse their young; and so on. But they also share an important feature, their streamlined shape. Any fast-moving, water-living organism must be streamlined to conserve energy. (Just imagine a cow-shaped torpedo.) Similarly, airborne animals need lift. Wings of birds, bats, and insects provide this lift. Bird and bat wings arose from transformed arms, and insect wings emerged from flaps of tissue on the leg or back. Despite their different origins, the wings of birds, bats, and insects have similar shapes because they operate according to the same aerodynamic principles. Solar-powered organisms need a large exterior surface—the solar panels we call leaves—to capture light effectively. Large land-living organisms need to build massive support scaffolds to prevent collapse, such as the imposing skeleton of elephants or the gargantuan trunks of redwoods.

The list of these predictable features could go on and on. Life's direction is sometimes clear when principles of physics are involved—hydrodynamics, optics, aerodynamics, or statics. But for any such example of the predictable, a devil's advocate could raise another one of the unexpected. The lifestyle of kangaroos and other marsupials evolved only in Australia. Why? And as the biologist John Maynard Smith once asked: Why do antelopes have horns and birds don't? We do not know.

Such is goal-oriented evolution as applied to individual lifestyles. It does not survive critical examination. What about cosmic teleology of life as a whole? Much of what we have learned since Darwin — for example, organisms' gamble with their genetic material or parasites' enormous success — has fueled disbelief about cosmic teleology in general. (For whom is a parasite's paradise a better world?) The many dead ends organisms have taken, all the lifestyle alternatives that never were, and the fleetingness of success all argue against final goals.

Not even the most obvious candidate directions of life survive scrutiny. Take the complexity of life itself: Is life becoming ever more complex, harboring more complex communities of more complex organisms? The naive answer is yes, because the story of rising complexity is apparently straightforward:[45] three billion years ago, simple one-celled life dominated the planet, then multicellular organisms arose, and then humans. But a close look uncovers many problems with this view.

First, complexity is a notoriously imprecise concept: every known approach of measuring life's complexity has flaws. For instance, some scientists have advocated "order" as a measure of complexity. They argue that order increases in the course of life's history. Yet not only are many of the living's parts quite unordered, but much order exists also in the inanimate world, in the form of crystals. Crystals are as ordered as they are dead, comprising billions of molecules arranged in a repetitive pattern. What, then, about simply calling the number of body parts or different cell types of an organism its complexity? But any large enough heap of junk might contain as many junk items as there are cells in your body. The length of a body's parts list is thus not a good measure of complexity either. And let's not forget that half of all life is parasitic. Parasites lose abilities and body parts — whether proteins or larger parts — because the host provides for them. Half of life may be becoming simpler as we speak.

No other measure — organization, integration, differentiation — has proven a more successful or universal indicator of complexity than order or the number of parts.[46] And if we cannot agree on how to measure complexity, how can we argue that it increases over time? Perhaps complexity is simply in the eye of the beholder.[47]

Life's Path toward History

Life's path does not look goal-oriented from near or afar. Perhaps it just meanders aimlessly through time? That is possible. But what if the quest for life's goal has distracted humans from an even simpler possibility? What if life's goal is the opposite of a goal?

What if life's path reduces — little by little — predictability, regularities, and the lawfulness exemplified by the shape of fish and the wings of birds? What if the historical, the serendipitous, and the individual become more and more important? The following argues in favor of this possibility. It also illustrates a pair of opposites related to risk and safety: *commitment and confinement* to a lifestyle and the new *opportunities and freedoms* such commitment can create.

Organisms that change and adopt a new way of life take chances. Flight is a departure from the beaten path of crawling and walking and thus carries risks. So do photosynthesis, life on land, and parasitism. The more radical a lifestyle change, the more it commits an organism to it. For tampering with it will often spell disaster. A sudden genetic change rendering an organism unable to live on land — say, by disabling its lungs or legs — will be disastrous. And what if an organism can no longer produce an important building block for growth, live just about anywhere, or digest just about any food? Extreme parasites provide the most striking examples of such loss of abilities, but any organisms, from free-living bacteria on, serve to make the point. Some bacteria are food specialists that need particular molecules, whereas others need special surroundings, such as an oxygen-rich environment. They are locked into particular ways of life that confine possible avenues of further change. They have become confined to their lifestyle.

You could look at such confinement as a prison from which there is no easy escape, no exit permitting further change. That is one side of the coin. The following analogy from human life illustrates the other side, the freedom such commitment can create.

Transistors and integrated circuits were milestone innovations of the twentieth century. They allowed the miniaturization of electronic computers, which have transformed human life as nothing has before them. Now we are utterly dependent on them. They permeate our lives, be it in mundane appliances we use every day (coffee makers, toasters, clocks) or in gigantic corporations and governments that rule us.

But electronic computing is not the only kind of computing.[48] Thousands of years ago, the Greeks used sophisticated mechanical computers to predict planetary constellations and daylight hours. In fact, before electronic computers were the rage, most computing relied on mechanical or electromechanical principles. Why didn't some of these technologies come to be dominant instead of electronics? The standard answer is, because miniaturizing mechanical computers is difficult. But is electronic computing intrinsically superior to *all* other technologies? No. Some alternative computing technologies use light, communicating molecules, even elementary particles. They allow even greater miniaturization. And what about the oldest and most powerful computers, computers using a conversation of cells, the computers we call brains?

Clearly, the ways to compute are legion. However, we are currently committed to one of them. The reasons for this commitment would be interesting to explore. Its history itself is complicated, including the development of a universal computer architecture — known as the von Neumann computer — operating system standards, programming languages, and communication protocols and technology. But instead of focusing on why we compute electronically, I want to focus on one consequence of this confinement, which is nothing short of a revolution.

Before this revolution, information — letters, documents, images — took weeks or months to travel around the globe. Since then, communication has become all but instantaneous. Instantaneous communication has not only facilitated global commerce. It is revolutionizing education, allowing those in remote regions to partake in an information economy; it is transforming the practice of medicine through telemedicine, allowing doctors to treat disease halfway around the globe; and it is changing countless other aspects of life, from how we shop to how we manage our time. A confinement to one way of computing has become a world of opportunity. From the walls of its prison grew the flowers of an unprecedented freedom to innovate and create.

An innovation like the electronic computer creates confinements, which generate new opportunities, which in turn generate new confinements, and so on.

Just the same holds for the lifestyles of organisms. The confinement of a species' lifestyle opens unprecedented opportunities. The first vertebrates to explore land spent a good part of their life in water. They hobbled onto land using fins that doubled as wobbly limbs. But once complete, their transition to land turned into a universe of innovation. It created snakes that expertly wind their way across rough terrain, leopards that chase their prey at breakneck speed, and monkeys that swing swiftly from tree to tree. It even paved the way for the conquest of air.

Solar power is another case in point. More than two billion years ago, an organism won a gamble and acquired the ability to exploit solar energy. Its success and the ensuing commitment triggered the evolution of plants, ranging from microscopic algae to immense blankets of rain forest covering a million square kilometers. And less revolutionary examples of innovations are all around us. Plants, for example, had to innovate to prevent animals from eating their body parts. Many plants release toxic chemicals. (Some animals learned to neutralize these toxins.) Other plants are thorny. (Some animals acquired leathery mouthparts to avoid injury.) Yet other plants armor their seeds. (Some animals learned to crack the armor.)

Life on different continents also illustrates how confinement to a lifestyle — living on the ground, in trees, or in water — can turn into a universe of opportunity. Land-living organisms on separate continents like Africa, the Americas, and Australia have lived separate lives for hundreds of million years. So have many water-living organisms around each continent. Separated by an abyss of deep water, these organisms have evolved separately. All these organisms have common features.[49] Fish are streamlined everywhere. Plants turn their leaves toward light. Transportation networks, whether they deliver food, oxygen, or water, are shaped like trees. But beyond such similarities lie many drastically different lifestyles unique to each continent. Popular examples are Australia's kangaroos and koalas and the armadillos of North and South America. Less familiar are the unique flying lifestyles of Borneo's rain forest, where tree snakes, frogs, and lizards set sail and glide through the air; the lifestyle of bloodsucking vampire bats, found only in the Americas; that of the cattle egret, a bird feeding on insects stirred up by grazing mammals, which originated only in the New World; that of sea snakes which inhabit only the Indian and Pacific Oceans. Each and every one of these organisms is living testimony that life can turn out in countless ways.[50]

Opportunity and confinement to a lifestyle are thus two sides of the same coin. Commitment and confinement create opportunities and the freedom that ensues. Freedom requires confinement, for freedom requires something to be free of. Conversely, to exercise the choices that freedom permits is to create commitment and confinement.

Notice that no commitment and confinement—whether human or nonhuman—is absolute. There is no prison without escape. Confinement to a lifestyle may favor some possible futures but never excludes any of them. This is obvious from examples—often exceptional, always illuminating—where commitments are reversed. Although most birds fly, some returned to walking; while most land-living animals stayed on land, some returned to the water; while most plants get their nitrogen from the soil, some get it from the air through nitrogen-fixing bacteria or from animals they eat.

Humans and their institutions are no exception. Countries whose constitutions commit them to democracy occasionally elect a future dictator. Corporations often commit to products and services—say, manufacturing machinery, accounting software, or building contractors—but reverse that commitment later at punishing costs.[51] And on the smallest scale, humans revoke commitments to other humans through broken contracts, divorces, or abandoned children. Commitment of one human to another creates freedom, and opens areas of life difficult to access alone. But it also confines "self" to a lifestyle shaped by some "other." As a result, self sometimes revokes its commitment, however absolute this commitment once appeared.

All in the Same Boat

You might argue that these parallels between human and nonhuman commitments are superficial. For the nature of the gamble differs greatly. People gamble by committing to other people, products, technologies, or ideas. Although a company's unsuccessful bet on a particular technology means bankruptcy, its employees usually survive. In contrast, an organism's unsuccessful gamble with its genes means death.

This difference surely seems fundamental. From a different perspective, however, it is the difference that is superficial and the similarities that are profound. This becomes clearer when considering the uncertain future shared by all gamblers for survival, whether societies, companies, or individual organisms.

Is the world safer, more predictable, and less risky than three billion years ago on a planet of single-celled organisms? Are the bets we make — the games we play with our future — safer than those made by a gambling bacterium whose genetic material changes at random? Consider that many of our gambles are even riskier than those of other organisms: eventually, they may not only cost our life but bring the entire planet down.

Take our lifestyle. Like any other organism, we need energy and produce waste, but we live longer, less strenuous, and healthier lives than any comparable animal. These benefits are largely afforded by technology. And they have a price: we need more energy and produce more waste than other organisms. Our life thus requires bets on particular technologies that can sustain it, and many of these technologies affect not only our life but that of the planet. We gamble on a much larger scale than bacteria.

Nuclear energy might free us from fossil fuels but could also poison the planet, not to speak of its dark kin, nuclear weapons. Reactor accidents and nuclear waste storage problems have largely turned public opinion against nuclear energy. But to determine nuclear energy's risks once and for all is impossible, as the many contradictory expert opinions on it show. Some call nuclear energy the only long-term solution to our energy problems, whereas for others it is the Devil incarnate. And what about transportation technologies like gasoline-powered cars? In light of the air pollution caused by them, switching to battery-electric or hybrid cars may seem an easy choice. But car batteries contain toxic heavy metals. Producing and recycling them consumes much energy, which itself needs to be produced and creates pollution. Once a billion electric cars circulate, the problem of pollution may simply shift from the car to the power plants that create its fuels. Or take the problem of building homes for billions of people. Wood is still a prime source of building material in some regions. How much forest can we cut down without tipping earth's ecological balance? And do mass-produced manmade building materials have no unforeseen side effects?

These are only some of the gambles influencing the fate of billions. Other high-stake gambles include collective choices of economic policies, governments, and even religions. In all these areas, as many opinions as people circulate, but no certain answers. In matters of long-term life and death, our position is thus similar to that of bacteria: the outcome of our gambles is unpredictable except in hindsight. A once-reasonable choice can lead to apocalyptic disaster. And whereas a bacterium's gamble can obliterate its life, we have the power to obliterate the entire planet. Life is as dangerous as ever.

It is through an unknown and unknowable future that safety and risk, though opposites, are intimately linked. Like two sides of a coin, they cannot exist without one another, for certain knowledge about the future does not exist. Questions about the risk of a choice do not have one true answer. The superficially safe may harbor great risks, because we evaluate its safety from the past, a poor guide to the future; the impossibly risky may provide the only refuge.

Because absolute safety and certainty have no part in it, this perspective is certainly unsettling. But it also buys us a world of freedom. It buys us dignity. It elevates us above machines. It gives us the freedom of an unknown future, the freedom to make choices, choices we make despite possibly terrifying consequences. To be sure, this future will confine us, confine us because of our past choices. But this confinement breeds freedom and opportunity. And a future of our choosing.

CHAPTER 5

Destructive Creation

Only the hand that erases can

write the true thing.

Meister Eckhart

A moment ago, I brushed another hair — another fallen leaf — off my writing desk. Lately I have been finding more and more of them. My hairbrush draws a richer harvest by the week. This tree is losing its leaves.

The analogy between falling leaves and hair loss is more than skin deep, because the same peculiar process drives both balding and a tree's preparation for winter. It is a process shaping all organisms, involving destruction as much as creation and death as much as birth.

Where a leaf attaches via its stalk to a tree branch, you can see a little knobby protrusion. This knob is the site of the bud from which the leaf emerged. It is also home to a thin strip of cells that is the glue attaching the leaf to its twig. In the spring, even a storm cannot pry leaves from their tree. In the fall, however, a gentle breeze, barely enough to cause a rustle, will sweep them to the ground. What changes as the season passes? The cells in this adhesive strip die. When few remain to hold the strip together, a leaf's own weight can break its fragile connection to the tree.

In contrast to leaves, hairs form in follicles, which are sheaths of tissue that protect the growing hair shaft and harbor the cells from which the hair emerges. Hair follicles undergo a three-stage life cycle: hair growth, regression (during which many follicle cells die), and a dormant stage (during which the hair is shed). After a few months in dormancy, a follicle begins to grow a new hair, and the cycle repeats. How long a follicle remains in growth determines the length of its hair: a hair follicle on the human scalp can nurture the same hair for years, whereas a follicle in an eyebrow begins to regress after mere months. Regression, the prelude for a hair's demise, involves the massive death of cells that glue the hair shaft to the follicle.

You might think that the cells in a leaf stalk or in a hair follicle simply die of old age. A reasonable thought — after all, nothing lasts forever. But why would only stalk or follicle cells die of old age? Why not the billions of other cells in a tree or a human body, some of which are much older? It turns out that leaf stalk and hair follicle cells do not die of old age. They commit suicide. Each cell in a leaf stalk is fine before it dies. But when its time has come, the cell itself activates a complicated machinery of proteins with the ferocious power of a bomb, which literally cuts the cell into pieces. Hair follicle cells use a similar machinery for their mass suicide. (A leaf dies when the tree prepares for winter, forcing me to think what falling hair foreshadows.)

Sculpting Life

Cell suicide is more commonly known as programmed cell death, or apoptosis. And indeed, the word "apoptosis" derives from the Greek word for falling leaves.[1] Cell suicide is as much a part of life as cell birth and cell division. Falling hair and dropping leaves might suggest that cell suicide occurs only when organisms shed parts or prepare for dormancy or, worse, for death. Nothing could be further from the truth.

Cell suicide is key wherever organisms form. To sculpt a statue is to chisel away at a block of raw marble, to eliminate most of the raw material. Similarly, to build an organism is to eliminate cells, often most of them. But nature's way of building bodies differs in one crucial respect from a sculptor's work, highlighting the unique elegance of creation: nature creates and destroys simultaneously. Both the making of the marble block and its sculpting — cell division that builds raw material and cell death that shapes it — occur at the same time, all the time. Further examples illustrate how this principle applies to all life.

Any large organism relies on sophisticated transport networks for survival. Trees are no exception. They ship water and minerals from roots to leaves, which use them to manufacture building materials. Like our network of blood vessels, a tree's transportation system consists of many hollow tubes that carry huge volumes of vital liquid.[2] Where do these hollows come from? Does the tree, as it grows, build walls of cells around hollow spaces? No. Early in a tree's life, these tubes are solid, filled with cells. As the trees grows from a tiny sapling, these cells die through cell suicide. Their sacrifice shapes the transportation network that allows large trees to exist.

When our hands formed in our mother's womb, the fingers were fused, joined by a dense layer of cells. At that time, our hands resembled spades or flippers. Eventually, though, the cells joining our fingers died. They committed suicide like the cells filling a tree's hollows. Had they not committed suicide, you would have been born with a malformation — syndactyly — recognizable through fused fingers. Without these cells' sacrifice, your hands would be as inflexible as the fins of a fish.

Our hands are flexible tools not only because our fingers are unfused. We must also be able to move these fingers, independently but in coordination. The same holds for any coordinated body movement. The muscles moving each body part must follow our instructions precisely. The material side of these instructions is electric currents traveling from our brain through nerves — living wires — to muscles. Each muscle cell, and there are millions of them, is attached to only one nerve cell. An electric signal traveling along a nerve instructs the attached muscle to shorten. That's about all a muscle does. But the coordinated shortening of thousands of muscle cells — a symphony of tiny twitches — allows dancers to defy gravity, yogis to contort their body into bizarre shapes, and musicians to express their deepest emotions. It allows my hands to write these lines and yours to turn these pages.

Like our hands, the living communication network permitting these marvels was created while we were still in our mother's womb. Back then, however, it was a tangled mess. Each muscle cell had many — not one — nerve cells attached. Each muscle cell thus received mixed messages: one nerve cell might instruct the muscle to shorten, while another might tell it to do nothing. In an adult, the result would be a body convulsing with uncontrollable muscle twitches.

Luckily, our bodies sorted out this mess around birth. How? Through the suicide of millions of nerve cells: all but one of the nerve cells attached to each muscle died. Their suicide prevents mixed messages, and is essential for controlled movement. The original tangle of living wires provided only the raw material for the sculpting of a sophisticated communication network.

This principle — chiseling away at abundant raw material — applies not only to the cells that allow you to act in the world but also to the cells that allow you to sense the world. Your eyes are a case in point. A dense network of light-sensitive cells and nerve cells in your retina allows you to read these lines. When the retina formed, about 90 percent of its cells committed suicide. The same holds for the myriad nerve cells communicating throughout your nervous system and in your brain. More than two-thirds of them — billions of cells — died in its formation.[3] Ultimately, their death allowed you to see the world around you, including this book. The science writer Lewis Thomas expressed this phenomenon as follows: "By the time I was born, more of me had died than survived."[4]

My last example, metamorphosis, a complete transformation in an organism's form and function, has fascinated people for centuries. Caterpillars that transform into butterflies illustrate it most dramatically, but metamorphosis occurs in innumerable organisms. Perhaps the best-studied metamorphosis is that of a tadpole changing into a frog, during which a tadpole adopts a land-living lifestyle and its body gets remodeled from the inside out. Most conspicuous is the disappearance of the tadpole's tail, which was essential for swimming but gives way to the frog's emerging legs. Less visible are many other transformations that occur simultaneously. For instance, tadpoles eat mostly plants, whereas frogs are predators. A frog's digestive tract is thus as different from a tadpole's as a carnivore's intestine is from that of a cow. The radical remodeling of the tadpole's innards takes place during metamorphosis. And a tadpole needs gills, which the frog sheds for fully developed lungs. Metamorphosis thus involves the death of millions of cells, and nowhere more than in the melting away of the tail, which accounts for half of the tadpole's body weight. Every one of the doomed cells commits suicide, is recycled, and transformed into new building material.

All these examples concern the making of an organism. But in another deviation from marble sculptures — which are unchanging once completed — many organisms remodel their bodies throughout life. Most of this remodeling, relentless but invisible to us, involves programmed cell death and eventual replacement of the dying cells by others. In your and my body, more than a million cells sacrifice life every day.[5]

To be sure, some of these cells may have become a danger to the whole. They may be on the verge of becoming cancerous, have suffered severe DNA damage, or become infected with a virus. Such diseased cells often remove themselves from life. But many others are perfectly normal, and their suicide is just part of maintaining a living sculpture. The scale of their self-imposed death is staggering: the weight of cells that die in your body every year adds up to your entire body weight.[6]

Deadly Conversations

How do cells commit suicide? And how do they know when their time has come? Like almost anything, cell suicide involves communication. Who or what communicates? The cells themselves. Most cells commit suicide because other cells tell them to do so.[7] This communication involves many molecules, but it is essentially the same as many other conversations we have encountered.[8] It is similar to the conversation leading to the sculpting of eyes, a conversation between developing brain and body surface. It is also similar to how food molecules and receptor proteins interact on the surface of bacteria.

As in these cases, the conversation begins with two molecules. One of the cells releases a molecular message from its surface — let's call it the kiss of death. The other cell carries a receptor protein, which you might call the suicide receptor. Similar to other receptors, the suicide receptor is part of the cell wall. It links the cell's interior to its exterior. Like a plug and its socket, kiss of death and suicide receptor have complementary shapes.[9] When they encounter, their complementarity allows them to connect, whereupon the suicide receptor changes shape. This shape change also transforms the receptor parts that protrude into the cell. Other proteins inside the cell recognize the ghastly meaning of this shape change. These other proteins change shape in response and become complementary to yet other proteins, which change shape in turn. And so forth.

This cascade of shape changes is what ultimately causes the cells to die, because some of the changed proteins become the knives that cut the cell apart. Their new shape endows them with a new surface, one that turns them into active enzymes. Each of these enzymes may carry out a different chemical reaction, but all the reactions have one thing in common: they destroy the cell's parts. Some enzymes cut other proteins into pieces; others slice into the cell's DNA; others demolish the cytoskeleton, the elaborate scaffold that gives the cell its shape; yet others lay waste to the membranes that separate the cell from the outside world.[10]

If you observed a dying cell under a microscope, you would see it literally fall into pieces, no matter how its suicide came about. An autonomous individual with a cell body and a clear boundary implodes into little droplets of debris enclosed by remnants of the cell wall. The process, beginning with the kiss of death and ending with the cell's demise, may take less than twenty minutes. Nearby cells usually ingest the remaining rubble and recycle it for future cell birth.

But is such cell death really suicide? After all, it usually requires communication among cells. To call apoptosis cell suicide is of course a matter of perspective. But consider that the dying cell *itself* harbors the destructive machinery that causes its demise. The other cells provide only a message that triggers the suicide.[11]

In sum, cell death of this sort — different from death through old age or through external damage — takes place countless times whenever the living sculpture of a body forms and maintains itself. Death is an integral part of creation. After having begun an unimaginable thousands of million years ago, it goes on to this day, every moment, wherever life occurs.

Creative Destruction Everywhere

If organisms are living sculptures of cells, then cells are living sculptures of molecules. And inside cells, creation and destruction also abound. For not only organisms and cells are born and die. Their parts are also constantly created and destroyed.

Take proteins, the most abundant cell parts. Each minute, a cell produces more than three thousand identical copies of some proteins, every one of them a complicated string of amino acids. If you consider that a cell contains many thousand *kinds* of proteins, the proteins it creates every minute must number in the millions. But they do not live forever. They do not even live very long by human standards: a long-lived protein may live for days, many others for mere minutes.

How does a protein's life end? A few proteins simply die a natural death, ceasing to function. But most proteins do not die from natural causes: other proteins, enzymes, cut them into pieces. (Unlike cells, then, most proteins do not self-destruct.)[12] Proteins, like cells, thus have a time to live and a time to die. And their lifespan may also be written into them, into the string of amino acids that endows them with identity. For some proteins, this message is encoded in the string's first amino acid. Long-lived proteins begin with amino acids such as alanine or proline; short-lived proteins begin with amino acids such as lysine or isoleucine.[13] These amino acids are signals to the proteins that destroy their kindred.

Thus, just like the countless cells in a body, the countless parts of a cell are constantly degraded and replaced. This seems like a terrible waste of effort. Why build millions of complex molecules every minute while simultaneously tearing down a million others?

Because the world changes constantly. Look at food and its changing supply. Swimming bacteria may get it from the surrounding liquid, plant roots from the surrounding soil, and your body cells from your bloodstream. But regardless of cell type, a cell's food supply constantly changes. A changing food supply requires changing proteins, changing enzymes that transform the new food molecules into energy. Cells thus destroy their old proteins and make new ones that recognize new food molecules. Even cells that have run out of food produce particular proteins, proteins allowing them to endure starvation for some time, perhaps by burning food stashed away for bad times.

A constantly changing food supply is but one of many aspects of a changing world. Temperatures around a cell also change perpetually, and heat is especially damaging to proteins because it can destroy their shape. In a hot environment, cells thus produce proteins that attach to others and repair any damage excessive heat may cause. Another case in point is toxic substances with potentially fatal effects that occasionally enter a cell's environment. In response, cells produce antidotes — often proteins — that neutralize these toxins. When exposed to radiation — even just sunlight — cells may suffer increased DNA damage, requiring DNA repair proteins to mend the harm. These and many other changes require a constant rejuvenation of a cell's content, because the limited space inside a cell needs to be filled with proteins specialized in the tasks any one environment requires.

To round out this image of a hierarchy of destructive creation, let's briefly revisit the chemical reactions that create and destroy the parts of the living, including proteins. They are carried out by enzymes, proteins that join two molecules, break them apart, or swap their parts. Life is sustained by billions of these chemical conversations, which are occurring every second in every cell of every organism. In each of these reactions, some molecule is destroyed and another one is created. The same holds for any chemical reaction, any at all, whether aided by enzymes or not, inside or outside organisms. It also holds for atoms joining to form molecules and for molecules falling apart into atoms. It holds when atoms change into other atoms, whether by taking in some elementary particle or by liberating another. It holds when elementary particles interact. In fact, elementary particles are the consummate experts at destruction and creation: destroying and re-creating one another is about all they do. As the physicist Richard Feynman facetiously stated, "Created and annihilated, created and annihilated, what a waste of time."[14]

The *Natural* Causes of Death

Destructive creation permeates a hierarchy of being from tissues on downward to cells and their smallest parts. It is thus little surprise to find it on the larger scale of whole organisms and their populations and communities. You and I are of course painfully aware that we do not live forever. However little consolation it may be, neither does any other organism. Even single-celled organisms grow old, cease to divide, and die.[15] Most of them, however, never get to this point. They starve to death, get eaten, burn to a crisp, or succumb to DNA damage long before dying from natural causes. The tiny minority that reaches old age raises an intriguing question: What are those *natural* causes that eventually kill all life, whether single-celled or many-celled? What is the *natural* life span of an organism? We do not yet know for certain, but several decades of biological research have narrowed down the possibilities.

Anything alive or dead that uses energy (cars, toasters, nuclear power plants, cells) produces waste. Such waste can be dealt with in different ways. Many-celled animals — like industrial facilities — expel most of their waste. Plants, in contrast, isolate waste inside their cells in membrane-enclosed spaces — a million tiny garbage pits inside their bodies.

Some cellular waste is so hazardous that it needs to be neutralized immediately. This waste consists of enormously reactive molecules, molecular suicide bombers that destroy any other molecule they encounter (and themselves with it). Organisms neutralize such waste before it ever leaves their body, the way a car's catalytic converter neutralizes the engine's exhaust. But despite an organism's efforts of mopping up — nobody is perfect — some of this dangerous waste damages cells. A certain amount of damage can be repaired, but after too much of it has piled up, a cell succumbs and dies. Accumulating cellular damage can also kill multicellular organisms. The natural death of an organism can thus be the result of its creeping self-poisoning through waste. And how long this self-destruction takes depends mostly on how much waste the organism produces, which depends on how much energy it burns.[16]

Energy and waste, however, are only part of the equation determining life span. The other part involves genes. Organisms within the same species (and thus with similar lifestyles) consume about the same amounts of energy and produce the same amounts of waste. They do not, however, live equally long. Why do we think that genes contribute to their different fates? First, some mutations, changes in particular genes, greatly delay or accelerate aging. Nowhere is this more striking than in rare human genetic diseases like Werner syndrome. Individuals suffering from Werner syndrome exhibit signs of aging by adolescence; they develop various aging-associated conditions such as atherosclerosis, diabetes, and heart disease early in life; many die in their thirties, with the bodies of centenarians. Other mutations that change life span are also known from such organisms as fruit flies and mice. Most of these mutations reduce longevity, but a select few increase it by more than 50 percent.[17]

A second reason to believe that genes influence life span is that longevity runs in families. If one family member is long-lived, its offspring also tends to be long-lived. In other words, life span is heritable. A third reason comes from laboratory evolution experiments in which only long-lived animals are allowed to reproduce. Such experiments can slowly increase longevity in fruit flies and other organisms. Their success underscores the role of genes in longevity.

Waste damage is thus not the only cause of natural death. The organism itself and its DNA play a role in it. But contrary to the clear purpose of cell suicide in multicellular organisms, the purpose of your and my death is less obvious. Why do we not live half as long or twice as long as we do?[18] Who does a short life span benefit, the organism, the population, or something else? Even though we do not have a final answer to this question, one thing is clear: death is no accident of life. Like the proteins and cells inside us, we harbor the source of our dying.

Death Benefits

In any population and community, organisms perpetually die. Each death of an organism benefits another, whether the predator that killed the organism, the neighboring tree that can now spread its canopy, or the bacterium, fungus, maggot, or vulture feeding on detritus or carrion. Other thrives through the destruction of self.

Another consequence of death is more important. Natural selection, I said earlier, proceeds through organisms with slightly different inherited lifestyles. An organism with a less successful lifestyle either dies early or leaves fewer offspring than others. If this less successful organism leaves fewer offspring, its family lineage — the family lifestyle — will eventually become extinct. The space it leaves behind can be filled by other organisms, new experiments, and innovations. Put differently, natural selection is impossible without death, without replacing ways of life with other ways of life. And although natural selection alone does not create the world of living things, the world would not be the same without it. Without natural selection and death, there would not be a world teeming with organisms, whether humans or submicroscopic bacteria. Without death, no multicellular organisms would have arisen. No entangled web of millions of species would exist, lifestyles nested within lifestyles, each depending on many others. None of the conversations shaping this web would take place either. Neither you nor I would have been born, for we are also the product of countless such conversations. I would not be around to write this, and you would not be around to read it. *Without death, earth would be as barren as the moon.*

While considering death *within* species, where slightly different lifestyles replace each other, let's not forget that the death *of* species dominates the history of life. The vast majority of species, highly successful when they originated, eventually perished. Whatever the reason for their demise — other lifestyles more cooperative or more competitive, gradual climate changes, sudden meteor strike or some other calamity — virtually all species eventually went extinct. And their survivors, the millions of species now alive, are only a tiny fraction of the ones that have vanished.

Living Worldviews

Similar principles apply to human life and culture, along with its customs, technologies, economies, governments, and scientific endeavors. Think of any way of life — human or otherwise — as built around theories, assumptions, conjectures, or hypotheses about the world. Each lifestyle is a strategy, a business plan, a secret for success built on these hypotheses. Take a parasitic plant whose roots — blindly but with unfailing accuracy — tap into a tree's root network to pilfer nutrients. This parasite *embodies* a theory about the world: that out there in the root network of trees are valuable nutrients; that to harvest them requires growing one's own roots toward this network; and that nothing, no defense mechanism, will thwart any such endeavor.

This is, of course, only the crudest description of a lifestyle. A careful assessment of parasitic plants would reveal countless more hypotheses about the world. Some parasitic plants only attack particular tree species, often identified through a signature chemical, a tree's smell. When catching a whiff of this smell, the parasite will grow its roots toward the tree. This behavior embodies another hypothesis: the source of the smell is a source of food. (The smell stands for the tree; it *means* the tree's presence.) And what happens as the parasite's subterranean feeding tubes have reached and begin to penetrate a tree's roots? In some species, the feeding tubes release a chemical loosening the tree's root tissue and opening a passage into the root. Behind this behavior lie specific hypotheses about the presence of certain molecules in the root, the chemistry of these molecules, and the released chemical's ability to alter them and allow penetration. When the tree replaces these molecules by others — perhaps as a mechanism of self-defense — this hypothesis may become false, rendering the chemical's release useless. Given enough time, we could dissect each facet of this lifestyle. Doing so would unravel an infinite number of assumptions about the world, the nature of other organisms, the laws of chemistry, and the most basic laws of physics.

The same holds for any other organism and its lifestyle. Any organism *embodies* innumerable hypotheses about the world. Social animals (animals that live in groups and cooperate, whether for hunting, self-defense, or rearing their offspring) embody the hypothesis that loners fare worse than group-living animals. A salmon swimming a thousand miles to its birthplace to spawn embodies the hypothesis that its birthplace is the best spawning ground. Giraffes' long necks embody hypotheses about plentiful food in tall trees. A bacterium swimming toward increasing concentrations of food embodies hypotheses about a food source, its distance, and its abundance. And a buzzing insect applies complex physical theories about aerodynamics and flight. If we looked closely enough, we would find countless further hypotheses, one built on top of the other, in each lifestyle. Organisms have been embodying these hypotheses since more than a billion years before our time.

From this perspective, organisms are living worldviews. Each of their lifestyles is a successful strategy based on a worldview, whether the strategy's utility is blatantly obvious, like that of a swimming bacterium, or mysterious, like that of salmon.

But aren't these assumptions and strategies different from the ideas, hypotheses, strategies, and business plans of people? For one thing, they are neither written on paper nor communicated in words. They do not live in the subtle world of thought but are carved in flesh. You could say that organisms *are* their ideas about the world. Thus, the organism itself answers the question whether its theories are flawed—by living or dying. A flawed theory means death.

In contrast, flawed human ideas, theories, and business plans usually do not spell death for a person. Human ideas and assumptions about the world (for example, those used for building steam engines, telescopes, or computers) are separate from steam engines, telescopes, and computers.

However, consider that the commonalities between worldviews are deeper than their superficial differences. First, the success of our hypotheses, like those of other organisms, hinges on detecting their flaws. An inventor with an idea for a revolutionary vacuum cleaner must build a prototype to find out whether the idea works. A company with a concept for an earth-shaking new product—whether a nuclear reactor or a sexier swimsuit—needs to find out whether the product works. The same holds for natural scientists and their hypotheses (for example, the hypothesis that plants turn their leaves toward light). Scientists test this hypothesis through experiment. They give the world a chance to prove them wrong.

A second fundamental similarity between human and other worldviews is that *something* dies with a flawed hypothesis, strategy, or idea. Whether hypotheses die together with the organism or not, creation requires them to die. (And keep in mind that human hypotheses—much like those of bacteria—also live in matter, be it in the firing of nerve cells or in the written word, both of which need a material carrier, however subtle.) Computing technology provides an apt illustration. Before the advent of electronic computing with transistors, computing technology used completely different principles: electromechanical relays, complex mechanical gears, the time-honored abacus, or pebbles in the sand. Electronic computing itself has not remained unchanged since its early days. It has come a long way from the early hulking computing monsters that filled a building and worked for mere minutes between countless hours of maintenance. Much of this change comes from improved ways of building transistors. We usually think of this change in terms of progress, because it allowed an annual near doubling of computing power that lasted for more than three decades. But progress is only part of the story. The other part is death, the death and abandonment of the old, whether mechanical computing, computing with vacuum tubes, or outmoded ways of building transistors. The road to faster computing is littered with casualties.

The same holds for every change in the human world: change requires destruction, whether or not you value the destroyed. This holds whether you consider drastic changes, such as in transportation (from horses to cars and airplanes), or subtle changes, such as those involved in building ever more efficient combustion engines. Not limited to technology, it also holds for the ever-changing ways in which people communicate, raise their children, cure diseases, and govern themselves. It holds for how humans build ideas and knowledge about the world or the ways of science. Wherever a new hypothesis, idea, or theory about the world becomes widely adopted, an old one fades away. And old ideas do not die easily. In the words of the quantum physicist Max Planck: "A new scientific truth does not triumph by convincing its opponents and making them see the light, but rather because its opponents eventually die."[19]

Death and creation always go hand in hand, whether it is the death of molecules, cells, organisms, species, and the worldviews they embody or the death of human concepts and ideas, as embodied in businesses, products, technologies, and scientific theories. Death and creation illustrate, once again, the two sides of our coin.

From their inseparability emerges a principle we have already encountered in different form: the ubiquity of death and destruction is the price to pay for the world's overflowing creativity.

CHAPTER 6

Choice in the Fabric of
Chance and Necessity

To know is not to know.

Not to know is to know.

Kena Upanishad

Can you make a difference in the world? If you lived a life such as that of Gandhi or Hitler, you would influence the fates of millions. Some of your decisions and choices would change human history. But can any human choice really change history's course? It might merely cause a temporary detour from a future deeply mysterious, yet utterly certain. No matter what we do, earth may always be full of war and misery; or it may eventually become free of war and misery; or we may eventually blow it to smithereens.

Here is a more modest question, a question that applies to world leaders and the rest of us alike: Are the choices you make your choices? Whether to go fight in a war, which cereal to have for breakfast, whom to marry, or when to watch a movie? Are these your own choices, or is the world around and inside you responsible for them, the world into which you were born, the family that raised you, or the genes your parents gave you? While you were reading this last paragraph, were you wondering about one of the questions? Was it your choice to wonder — perhaps to wonder who chooses to wonder? Or are you a cog in a monstrous machine that runs your life, one whose goals — if any — are far beyond your grasp?

Philosophers have argued about these questions for centuries and with remarkably little progress. Consider that this failure has a reason. To see it, we must explore the relation between chance and necessity. After that, we will be ready for a perspective that extends far beyond the human realm.

Castles Built of Sand . . .

Do you remember how bacteria swim for a living? If their corkscrew-like flagella turn counterclockwise, they swim straight; if the flagella turn clockwise, they tumble. Each flagellum changes directions erratically. That means we cannot predict how much time will pass between two successive changes of direction. We can, however, predict what really matters, that the bacterium will arrive at its destination, a food source. After a bacterium has picked up the scent of food and starts to swim in one direction, its flagella change directions less often if the scent gets stronger. If the scent gets weaker, its flagella change direction and reorient the bacterium more often, thus ensuring that the bacterium will eventually find the food source, despite all its erratic tumbling.

Such predictability is the basis for laws of nature:[1] everything else being equal, whenever A takes place — put food near bacterium — then B — bacterium swims toward food — will occur. The predictable behavior of a swimming bacterium results from many individual events, each of them utterly unpredictable. Predictability and lawful behavior emerge from unpredictability and chance.

The bacterium is not unusual in this regard. Scientists have known for a long time that laws of nature are founded on this principle. But we often forget how pervasive it is.

Recall how eyes form. Their creation requires communication between two groups of cells, the neural tube and the ectoderm. Neural tube cells release a molecular message that receptor proteins on epidermal cells can recognize. In response, the ectoderm forms a tiny pit on the body surface. After the exchange of innumerable further messages, eye parts such as lens, iris, vitreous body, and cornea develop. In every human and in many other animals, eyes form in this way. Call it a law of embryonic development: that eyes begin to form whenever the neural tube sends the ectoderm a certain chemical signal. Similar laws govern the conversations that form every body part, from your fingernails to your brain. But where is the unpredictable basis of these laws?

Imagine you want to cross a wide avenue packed with thousands of people, perhaps during a political demonstration. As soon as you enter the crowd you realize that any effort to cross will be futile. You are stuck within the moving mass of people, shoved this way and that, a powerless particle. If you wait long enough, the crowd may spit you out at the other side, but who knows when and where?

A single molecule released by the neural tube finds itself in a similar situation. It floats in the space between two tissues, surrounded by billions of other molecules. All these molecules bounce around erratically. No one could predict the moment when any two particles will bounce into each other or the force of their collision.

What makes these particles bounce is heat. Thermodynamics tells us that heat *is* particle motion.[2] You can visualize such motion indirectly by placing a tiny drop of ink into a perfectly still glass of water. Even without your stirring, the innumerable ink particles will eventually spread through the glass through the pushing and shoving of the invisible water molecules.[3] The molecular message released by the neural tube is just like one of these ink particles. It moves erratically this way and that.

Despite all this erratic motion, cellular conversations have predictable outcomes like eyes and brains because the neural tube releases not one but countless many molecules, each containing the same message. Just like dye particles in a glass of water and humans in a densely packed crowd, these neural tube messages are pushed around haphazardly. Timing and strength of each push are unpredictable. Equally unpredictable is whether a particular molecule will reach the ectoderm, how long its journey might take, and whether it will bump into its receptor. Despite these uncertainties, some of these molecular messages will arrive and latch on to their receptor simply because the neural tube released so many of them.

The receptor then changes shape and thus initiates another communication process, this time among other molecules bouncing around inside an ectodermal cell. This communication process is based on the same principles, the pushes and shoves of a myriad molecules, each utterly unpredictable. Many layers of molecular conversations later, the ectoderm will have reliably formed a lens.

The precision and predictability with which tissues, organs, and organisms form in this manner is incredible — but not absolute. We think of laws of nature as inviolable rules that govern the universe, but knowing the origin of developmental laws gives us pause. Could the neural tube fail to make the ectoderm form a lens, say, because all of its million messages happen to be pushed the wrong way, and none arrives? Yes, that is possible, as possible — and as unlikely — as getting "heads" every time in a million coin tosses. Probability theory, the mathematics of chance events, tells us that the odds of being born blind for this reason are very small. But not zero. The laws of nature are not inviolable. They do have tiny loopholes.

The predictable also emerges from the unpredictable on smaller scales. Consider, for example, the chemical reactions that transform molecules and atoms. If we observe a test tube containing two kinds of molecules, A and B, we cannot predict whether any two of these molecules will encounter each other. And even if they did, the outcome, whether they react to form C or D, is often equally unpredictable.[4] Predictable is only what A and B will do on average. Put a billion molecules of A and B in a test tube, then on average a certain number of encounters will result in C and the remainder in D.[5]

On an even smaller scale, the laws of quantum physics may prescribe the outcome of encounters among elementary particles more rigorously. But in the quantum world, Heisenberg's well-established uncertainty principle makes the encounter itself unpredictable, even more so than for molecules: measuring a particle's position accurately leaves its momentum unknown, and vice versa. However, we need to know both position and momentum to predict a particle's trajectory and thus its encounter with others. This uncertainty does not reflect a limitation of the human mind. It is a fundamental principle of physics. The macroscopic laws of physics thus also emerge from the average behavior of innumerable, microscopic, and unpredictable parts.

All these examples involve small-scale phenomena, such as elementary particles, molecules, and bacteria. Much larger objects can behave just as unpredictably yet produce lawful behavior. For example, the nerve cells in an animal's brain communicate through electrical signals. One nerve cell fires off a signal that travels to another nerve cell. When the signal reaches the other cell, that cell may fire as well, triggering the firing of a third cell, and so on, until innumerable cells relay signals back and forth. Whether any one cell will fire at any given moment is unpredictable. However, the result of many nerve cells conversing in this way can be extraordinary precision and predictability.[6] A skydiving raptor grasps its prey mere centimeters before smashing into the ground; a frog hurls its projectile tongue with great precision to catch the tiniest insect; a monkey, one of the forest's acrobats, swings through its canopies and never misses a branch. (It had better not miss, because a wrong move could mean death.)

All these marvels come from a brain's exquisitely precise calculations, based on the firing of innumerable nerve cells. Each cell's firing may be as unpredictable as a bacterium's change in swimming direction. But the resulting calculation is so precise that every animal, in every moment of its existence, bets its life on it. Think of this precision the next time you cross a busy street and navigate through a moving maze of passing cars.

Societies — human or otherwise — are not exempt from these principles. The example of racial segregation is a case in point. In the past fifty years, the U.S. government has made great strides to promote and enforce racial integration. Yet racial segregation persists, for example, in the segregation of residential neighborhoods. The economist and Nobel Laureate Thomas Schelling showed that such racial segregation does not require anything as crude as racism.[7] It can emerge in an utterly predictable and lawful manner from many unpredictable and subtle human decisions.

Imagine a densely populated urban area in which racial segregation does not exist. The homes of whites and nonwhites, for example, may be randomly scattered. In some areas, there might be slightly more whites, and in others there might be slightly more nonwhites. These differences might exist by pure chance, as when you scatter salt and pepper on a plate, and some areas contain more of one or the other. However, nonwhites might have an ever so slight preference to live in neighborhoods with many other nonwhites; whites might prefer neighborhoods with many other whites. This does not mean that nonwhites dislike whites, and vice versa. Just like you might unconsciously spend more time with people of like interests and value, so might the residents of this city. In addition, these preferences need not be strong. They might be hardly noticeable.

Families occasionally move within this city, although when and where may be completely unpredictable. Each family, because of its slight preference, may tend to move into an area with more residents of its own race. Schelling showed that over a long period and after many individual moves, nearly complete racial segregation emerges from such slight preferences. A random salt-and-pepper pattern of residences becomes completely segregated into white and nonwhite neighborhoods — despite the random initial distribution of neighborhoods and despite the unpredictability of each family's move.

Another example regards the spreading of such infectious diseases as the plague, measles, and smallpox. Your body might harbor, unknown to you or others, a disease-causing bacterium or virus. If you and I met, nobody could predict whether you would infect me or whether I would become ill or spread the disease to others.

From the vantage point of a government needing to preserve public health, the encounter of two people does not matter very much. What matters is to prevent epidemics, for instance, by vaccination. It is impossible to vaccinate everybody in a population: some people live in remote areas, others may refuse to get vaccinated. But vaccinate too few people and the remaining susceptible population will spread the disease.

What percentage of a population must be vaccinated to prevent an epidemic — 50 percent, 80 percent, or an impossible 100 percent? Individual encounters among people and their outcomes are unpredictable — they are, after all, based on individual choices — but it may be possible to estimate, *on average,* how easily a disease spreads from person to person and how many further infections a diseased person causes. Such information can suffice to predict what percentage of a population must be vaccinated.[8] Nearly eradicated diseases — smallpox and polio, for example — speak to the success of vaccination campaigns. This success is based on regularities of disease spreading.

In these and many more examples, lawful behavior derives from unpredictable events on the scale of whole organisms. Like any law, the laws behind such behavior have loopholes — some bigger, others smaller. Some loopholes result from unpredictable choices of humans, such as the choice of where to live and whom to socialize with.[9]

. . . and Sand Built of Castles

Rules and laws govern phenomena such as chemical reactions, embryonic development, and human societies. These laws emerge from unpredictability at a smaller scale, from atoms to people. But emerging laws are only one side of this coin. The other side is emerging unpredictability, emerging chance.

A time-honored law of physics, Newton's law of gravitation, provides the perfect illustration. According to Newton's law, two bodies attract each other with a force proportional to their mass and inversely proportional to the square of their distance. This is why objects fall to the ground and why the earth revolves around the sun. Newton's law makes predictions about many phenomena all around us, including the velocity of a falling body, the length of the lunar cycle, and the lift necessary to keep a plane in the air. Innumerable measurements have validated these predictions. If there ever was a law of nature, Newton's law is it. Yet this law can breed the utterly unpredictable.

Newton's law sometimes allows us to predict the future with great precision — for the moon rotating around the earth or for two suns spinning around each other. Knowing position and speed of two celestial bodies, it can predict their position and speed one year, a thousand years, or a million years into the future. But just take *three* celestial bodies, an entire solar system, or — heaven forbid — the millions of suns in a cluster of stars. Then predictability falls apart.[10] Newton's law itself is not violated, mind you: we can calculate precisely how much force each body exerts on every other. We can know everything there is to know about these bodies' attraction. Nevertheless, we are unable to predict the positions of only three bodies into the distant future. This particular type of unpredictability — for all intents and purposes, an indeterminate future[11] — is often called deterministic chaos.[12]

To understand the mathematics of deterministic chaos is not essential for our purpose, except for one point. We could predict the indefinite future of three celestial bodies if we knew their position and velocity with infinite precision. But if we know position and velocity only to some degree of accuracy, we can predict their future for only a limited time, say, a year. After two years, their whereabouts would be unknown. With tenfold greater accurate measurements, we might be able to predict their future for two years, but after four years we would again be left in the dark.

But is the erratic behavior of three bodies then *really* erratic? One might argue that we can predict their future *in principle,* simply because a law of nature completely determines their future *if* you have precise information on the present.

Most laws of nature, especially laws of physics, are built around infinitesimal calculus. When applying infinitesimal calculus to solve a problem, one pretends to change a variable (think: position of a planet) by an infinitely small amount and then studies how other variables (think: force of attraction) change in response. This line of reasoning works extremely well in practice. However, we have no way to observe, measure, or generate infinitely small changes in anything.[13] Infinitesimal calculus is a game of pretense, however powerful. The danger lies in forgetting that and confounding our pretense with reality. Only then are we tempted to say that we can predict the motion of three bodies in principle.

Even a million years from now, we will not know anything with infinite precision, nor will any finite mind in this universe. I did not even have to resort to quantum physics in this argument, which prohibits — in principle and practice — simultaneous and arbitrary precise measurements of speed and velocity of any object. In sum, the future of three bodies dancing around one another truly is indeterminate, even if you know everything one can know about them.[14]

The French mathematician Henri Poincaré was the first to uncover deterministic chaos, but it remained obscure until the middle of the twentieth century. It suddenly reappeared on the stage of science when several pioneers independently discovered chaos in areas of science as different as chemistry and ecology. Among them was Edward Lorenz, an MIT meteorologist who studied chaos in weather. He simulated weather on a computer using a few simple physical laws, such as those predicting wind speed from air pressure and temperature. The laws going into these simulations were as rigorous as Newton's law of gravitation. Yet the outcome was the constantly changing, capricious, and unpredictable behavior we expect from weather.

Chemistry experienced a similarly unexpected rise of the unpredictable. Consider a chemical reaction that predictably transforms two molecules, A and B, into C and D. The concentration of A and B in a test tube that may contain billions of each molecule will approach zero as C and D form and rise in concentration. Even simple chains of a few chemical reactions, however, can eradicate such predictability. For example: a chemical reaction transforms A into B, another reaction changes A and B into three molecules of B, and a third forms C from B. C itself might then react with yet another molecule D to recover A. Each reaction's outcome may be known precisely. But if all these reactions occur in the same place—whether in a test tube, a bacterium, or the human body—the concentration of A, B, C, and D may change completely erratically. The chain behaves chaotically, even though the laws of chemistry govern its links.[15]

Not only such tiny things as molecules and huge things as planets can behave chaotically. Things of intermediate size can, too. Take organisms and their populations as an example, and consider predators and their prey, such as snakes and mice or foxes and rabbits. Wildlife management aims to control populations of such organisms. Simple ecological laws that rule predator and prey populations can aid in this task. They may predict the average number of prey a predator consumes, a predator's life expectancy, or how fast prey populations grow when left alone. However, even though such laws may tell everything about how predators and prey influence each other, it may be utterly impossible to predict their population changes over many years. Like the chaotic dance of planets and molecules, societies of the living can show deterministic chaos.[16]

These are only a few among innumerable examples in which laws of nature breed the unpredictable. The location of a fly in your coffee as you stir it, the daily ups and downs of stock prices, the white blood cell count in leukemia patients, the swirling of white water in rapids, the exact place where a gently tumbling leaf will hit the ground—none are predictable, even though laws of nature may govern them.

All of the above goes to show that the erratic breeds the lawful and vice versa. Laws emerge from chance, and chance reemerges from laws. We find this principle everywhere from atoms to galaxies. It is not restricted to any part of the world. The large is no more predictable than the small, the dead no more than the living.

The predictable and the unpredictable are interwoven wherever you look. A perfectly ordered crystal can grow from the same liquid hosting a chaotic reaction's erratic dance; this chaos, in turn, itself emerged from chemical laws. From physics to sociology, from living cells to star clusters, chance and law are two fibers that form a tightly woven fabric. Each fiber of this fabric, examined carefully, consists of countless more fibers, each of which is part of an even finer fabric of law and chance. And so forth, unendingly. There is no center, no beginning, and no end of this fabric.[17]

The Choice to Choose

How does all this relate to the question whether your choices make a difference? It shows that no human can provide one final and universal answer to this question. You can focus on any one fiber of this fabric — for example, the center of your world, yourself — and examine the threads of chance and necessity running through it. The answer depends on which fibers you choose to focus on. To use an earlier analogy, you can focus on a molecule, bouncing around erratically in a liquid, or at the regular shape of the crystal it might become attached to. In other words, nobody other than you can answer this question.

This is nothing but a perspective. Why choose it? Because it gives humans — and the living — the power to choose their future and that of the world around them.[18] But the more powerful the perspective, the higher its price. This one requires us to give up certainty. (Paradoxically, this certainty includes the *certain* knowledge that your choices will make a difference.)[19]

The Seeds of Choice

Where are the roots of our choices? To get at this question, let's revisit the swimming bacterium and some other organisms. The bacterium's combination of straight swimming and erratic tumbling may seem an odd way of swimming for a living. But recall that if you were in a similar situation — blindly navigating a three-dimensional water world — you might search in much the same way: move in an arbitrary direction; if the smell gets stronger, stay the course; if not, turn toward a new direction. Most would say that your choices are responsible for your erratic trajectory. But how does it come to these choices? We don't know the answer for humans, but a look back at swimming bacteria will be instructive.[20]

We saw earlier that bacterial swimming involves a complicated conversation among food molecules, receptor proteins for them, and proteins that signal the swimming direction to the engine turning the flagellar propeller. Exactly when the engine will switch from forward to reverse rotation — by changing shape — is unpredictable. Part of the reason lies in the millions of molecules that surround the engine. All these molecules bounce around erratically in an incessant pushing and shoving and bumping. The other part of the answer lies in CheY-P, a molecule we encountered earlier. Recall that when this molecule attaches to the engine, the engine changes from counterclockwise to clockwise rotation, from straight swimming to tumbling. After a while, the bouncing of other molecules will tear CheY-P away from the engine, and the direction reverses again. I said earlier that CheY-P forms elsewhere in the bacterium—at the food receptor—and in response to the cell's swimming direction. When the cell is on the right track, it makes very little CheY-P, but when on the wrong track, it makes more and more. And to see why we cannot predict changes in direction, we have only to join these two pieces to answer one question: How does CheY-P travel from food receptor to engine? CheY-P, like other molecular messages we have encountered, can get to places only through the erratic shoving of its neighbors. If a bacterium contains very few CheY-P molecules, none of them may ever get pushed into the engine. But if the bacterium contains sufficiently many CheY-P molecules, at least some will reach the engine quickly and switch its direction.[21] When exactly they will arrive and how fast the erratic mob of neighbors will rip them away again nobody can know.

All of this molecular drama is hidden from us. The average microscope could reveal neither bouncing molecules nor CheY-P's erratic journey.[22] All it can reveal is a bacterium swimming this way and that and eventually reaching its food.

Does the bacterium *choose* to change direction? Somebody who sees, for the first time, a swimming bacterium might be tempted to say yes. But many who know what you now know about swimming bacteria might say no. After all, it is just bouncing molecules that cause the change in swimming directions. Decisions and choices, they might say, are the privilege of organisms like us, many-celled organisms with a complicated nervous system, consciousness, self-awareness, and free will. Before hastily leaning toward one of these positions, let us look at further examples.

For honeybees, collecting food is a life-and-death matter. Every day, a colony sends many bees to forage for nectar and pollen. Each forager may search many kilometers from the colony for flower patches to harvest nectar and pollen. But not all flower patches are created equal. Some are vast and rich and could support hundreds of foraging bees. Others are small and may have been overgrazed by other insects. Flooding them with foragers would be a waste of time. Not surprisingly, bee colonies choose among flower patches, dispatching more bees toward richer patches. A colony's choices are flexible: if the quality of a patch decreases, a colony quickly pulls workers from it and assigns them to a new patch.

Foraging bees that have returned from a rich flower patch perform a famous waggle dance.[23] This dance is rife with information about direction, distance, and richness of a patch. Its meaning is absorbed by "unemployed" foragers in the hive. Each watches a dance and then either goes out to the described patch or remains in the hive. After their dance, the returning bees usually depart again for the same patch. Foraging bees coming back from a poor patch, however, may decide not to dance at all and never return to this patch. This exchange of information, a process of self-organization, is simple in principle, but it involves the decisions of hundreds of foragers that form the foundation of the colony's decisions. Most important is that no one bee decides for the colony, because the colony's decisions are collective. The results speak for themselves: a single colony can reap more than 120 kilos of nectar per season.

Ant colonies of the species *Leptothorax albipennis* do not build elaborate nests but inhabit flat crevices or tiny rock caves. A colony that is ready to move and has found a new cave faces a key question: is the new property spacious enough to accommodate the family? The colony makes a decision. It stays out of caves that are too small and will move only into a sufficiently large one. It arrives at this decision in ways similar to how honeybees exploit flower patches, except that the decision is all or none. Individual scout workers go out to explore caves. They measure the size of each cave and return to the colony, which collects their findings and decides.[24]

Many social organisms make even more complex collective decisions, including those of fish schools that fluidly change their shape to evade predators, those of termites that build baroque castles, and those of some insects that regulate colony temperature with exquisite precision.[25]

Choice and Unpredictability

There are many differences between human behavior and the nonhuman behaviors we just encountered. Humans may have the most complex network of electrically conversing cells and the most refined cognitive abilities. Self-servingly, we like to think that human awareness and free will are necessary for choice (even though millennia of thinking about these concepts have not led us to agree on what they mean). Consider, however, that beyond these differences lie two more profound commonalities.

First, even though we do not know the molecular roots of human behavior as we do for swimming bacteria, these roots surely exist. Somewhere in our body, events exactly analogous to CheY-P bumping into the engine surely take place whenever we make a choice.

Second, human and nonhuman behaviors have another key commonality: unpredictability, a signature of choice.[26] Unpredictability indicates choice wherever we cannot ask an individual whether it chose. Take any of the choices people make every day, such as what to eat, what to wear, where to take a walk, or when to go to sleep. To everybody but the chooser, they are usually unpredictable.

Can choice then be reduced to some seemingly trivial unpredictable event, such as one molecule, CheY-P, bumping into another? If so, many would be tempted to reject this perspective. But recall that when I first mentioned swimming whole bacteria and their molecular parts, the question arose which of many proteins—food molecule, receptor, CheY-A, CheY-P—lets the engine know whether the bacterium is on the right track. We saw that you could pick any of them and maintain that it is key. Equally validly, you could argue that all of them together, as a whole, are what matters. Does the forceful bumping of CheY-P against the engine cause a shift direction? Or is it the shove of its neighbor, the shove before that, or any one of the myriad bumps jittering CheY-P across a cell? Whether to call any of these parts or the whole responsible for swimming is up to you. From this vantage point, the molecular roots of choice are no longer simple.

Some people might dislike the materialistic flavor of connecting choice to molecules. Choice involves nothing but bouncing molecules, matter? This perspective seems to lack basic ingredients of choice, ingredients such as mind and meaning. But recall that you can view every event in the world—especially if it involves the living—as being part of a conversation rife with meaning. A food molecule bumping into its receptor means something to the receptor. So does a tissue sending a molecular signal to another, the nerve cell instructing a muscle to shorten, and the billions of other molecular conversations that shape and sustain an organism. Each is part of a giant conversation, the monumental epic we call a body. To call this perspective materialistic thus reveals nothing but our preferences for one side of the coin.[27]

In sum, despite profound differences, human and nonhuman behaviors share key features. Do these commonalities mean that bacteria, insects, or their colonies truly choose? Only you can decide, for this is not a matter of truth but a choice of perspective, no more.

Before moving on, let me mention that the regular, predictable, and lawful can emerge from the unpredictable also here. Each change of the bacterium's swimming direction may be unpredictable. But the bacterium will predictably arrive at a food source. Just when any one foraging bee leaves the hive is unpredictable. Yet the colony will exploit its food sources optimally. Exactly when any of your brain cells fires as you cross a street is unpredictable. Yet you will most likely cross the street unharmed. All the behaviors I mentioned — to swim toward food, to forage at rich flower patches, to select a large cave — have something in common. They are for the best of the one that does the choosing, whether it is the bacterium, colony, or swarm. They are predictable. And they emerge from the unpredictable at a smaller scale, such as the individual cell in the brain and the insect in the colony. Of course, some bacteria may occasionally swim away from food, and some fish swarms may not evade a predator. Such behavior, however, will be short-lived. We will rarely find it.[28]

Life in a Web of Chance and Necessity

On the surface, the next question has little to do with chance and predictability. Why are most of us blind to the paradoxical tensions we have encountered here and that immerse us? Why do we have an urge to ignore them or to explain them away? The answer may lie in how life copes with the web of chance and necessity all around it.

Recall first that we can view an organism's lifestyle and behavior as representing a worldview that consists of convictions, assumptions, or beliefs about the world. Organisms and their actions *embody* these predictions. A plant turning its leaves toward the sun lives a prediction: light is a source of energy. A virus infecting a cell "predicts" that the cell will reproduce its viral genes. A bacterium swimming this way and that, a parasitic weed invading other plants, and a predator chasing its prey all embody the same prediction: that their object of desire harbors food. When a newborn chick sees a moving shadow overhead, whether from a falcon, an airplane, or a piece of cardboard, it cowers and hides. Male giant tortoises attempt to mount any object similar to a female, including other males and large rocks.

Organisms predict the world in which they live.[29] Their survival completely depends on the accuracy of these predictions, and unlike human predictions, they do not live in the subtle world of thoughts. They live in the coarse world of flesh.

Organisms have arrived at successful predictions through a string of trial and error lasting millions of generations and billions of years. Errors spell death, and we therefore see only the results of successful trials. For predictions embodied in animal behavior and cognition, this perspective was first made prominent by the philosopher Karl Popper and the biologist Konrad Lorenz.[30] As the following examples illustrate, it extends to all features of living organisms.

Successful life does not bank on the unpredictable. Organisms that decompose a forest floor rely on dead organic matter — their food source — reliably accumulating over the years. They would be doomed if they relied on one kind of leaf falling at a precise time. Many plants make modest weather forecasts, counting on a few rainy days early in a season to germinate. A plant that relies on rainfall on a specific day would be doomed in most climates.

Living organisms thus extract information on what is predictable, lawful, and orderly about the world. To live a prediction is to have "discovered" or "extracted" a law of nature, and the range of laws life has discovered is truly stunning. Many are laws of physics. For instance, where does the fuel come from that moves a bacterium's flagellum? Recall how dye drops spread from one part of a glass filled with water through the rest. The flagellum uses the same principle to generate energy. The flagellum is housed in a membrane, a wall, on one side of which the cell amasses innumerable hydrogen ions. When the bacterium opens holes in this wall, the ions move from one side to the other, by the same principle that causes ink particles to spread in water. In doing so they carry out work that cranks the engine, like water falling onto a turbine.[31]

Many other laws of physics are embodied in the large surface areas of wings, the streamlined shape of fishes, the treelike filigree of our blood vessels, the malleable eye lens, and the rigidity of bones.[32]

Life embodies not only physical but also chemical laws. It does so in the thousands of chemical reactions that sustain every living cell. A different enzyme facilitates each of them. And each enzyme, with its exquisite shape match to a molecule, stands for a chemical law that permits a particular reaction. This chemical ingenuity becomes even more remarkable, considering that life has learned to put just about every imaginable molecule to some use. Life can thrive on foods as bizarre or toxic—to us—as petroleum, benzene, heavy metals, hydrogen sulfide, ammonium, or barren rock.[33]

Finally, organisms also embody predictable features of other organisms, as I highlighted earlier. Every time a caterpillar grazes on a leaf, a parasitoid injects its eggs into the same caterpillar, or a bird eats the parasitoid, predictions are involved.

Some of these predictions about life itself may seem trivial and self-evidently true. But beware. Even the most self-evidently true worldviews fail. Cells often defeat viral predictions by chopping viral DNA into little pieces; a predator may chase its prey only to find it poisonous; immune systems, defenses of self against other, serve only one purpose: to defeat an attacker's predictions. In addition, each law of nature has limitations. Take social animals—from bees living in a hive to humans. They embody the law that cooperation is better than competition. But the success of many solitary animals, from flatworms to hedgehogs, belies the universality of this law. And what about the rapidly dividing cells in a growing bacterial colony? They embody a prediction about what matters in life, a prediction very different from that made by the millions of cells in our body, cells ready to sacrifice their life during every moment of our existence.[34]

The Evolution of Metaphysics

Among all the predictions that allowed our ancestors to survive, one class of predictions is especially important for my purpose. It reflects laws so fundamental that they may form the foundations of philosophy and metaphysics. Like a fish is unaware of the water it swims in, we are probably unaware of most of these predictions. Precisely for this reason, they hold us in an invisible iron grip.

Even infants younger than six months already make some of these predictions. For instance, infants already assume that any moving object around them is a whole that has a concrete shape, unchanging as it moves. This is no small feat, because part of the object may be concealed by another object, or the object may rotate as it moves and thus change its appearance every moment, like a toy thrown across a room. Infants also assume that moving things don't move in jumps but that they follow some more-or-less smooth, continuous curve. And they predict that two things can influence each other only when they touch. These are very profound predictions about the world. They resemble metaphysical laws such as that prohibiting action at a distance.[35]

Where do such fundamental predictions come from? Consider that they originated in evolution, through many hundred million years of successful and failed predictions about the world. According to this perspective, living things gradually acquired such metaphysical assumptions — resulting in unquestioned and unquestionable perspectives about the world — because they are useful. But these laws are far from inviolable. Some objects change shape when thrown across a room — just think of a length of rope. In these cases, such metaphysical assumptions can fool us, much as a female-shaped rock fools the male tortoise.[36] However, by and large these predictions must have been reliable enough for millions of life-and-death decisions, like those a monkey makes every instant on its acrobatic journey through the forest.

Earlier, I examined some of our ways of viewing the world that are no less fundamental than these metaphysical assumptions. I examined how we separate self and other and how we categorize anything and everything, such as the living into biological species. I examined the distinctions we make between creation and destruction, between parts and whole, and the way some of us raise one over the other. Whichever stone you turn, you will find more such distinctions, endlessly. We do not know for sure whether they originated in evolution, and perhaps we never will. But consider that these ways of carving up the world work so well, and are so self-evident, that we do not recognize them as painfully acquired by our distant ancestors, all the way back to bacteria.

To return to the example of self and other, their fates appear separate, but examination often reveals their inextricable connection.[37] Similarly, our obsession with categorizing objects sometimes fails disastrously, as in the earlier example of ring species with their overlapping populations. When confronted with these and many other failures, we tend to argue endlessly about the one *true* way of categorizing objects like organisms. However, the problem may not lie in the choice of category. It may lie in an unquestioned but violated metaphysical law we follow, predicting that the world can be completely organized. What if such failures carry a lesson? What if these apparently microscopic defects in laws of nature turn out to be gaping holes when examined closely? What if they mean that the world does not cater to our yearning to understand, that many of the paradoxical tensions we encountered are fundamental to it?

Consider that, ultimately, the foundation of all unresolved philosophical disputes dominating Western philosophy over two thousand years is just this: the unshakeable — because tacit — conviction that the world must be completely understandable, combined with the paradoxical nature of the world's foundation. Neither party to a dispute sees that its entrenched position is just one side of a coin. It raises self over other, whole over part, body over mind, matter over meaning, or vice versa and denies the other side legitimacy. Nothing may be wrong, except the belief that one of them is the truth. This belief is an outgrowth of a law all organisms live by, for we all live — and die — by our predictions.

CHAPTER 7

Purposeful Openness

Our imagination is the only limit to what
we can hope to have in the future.

Charles Franklin Kettering,
inventor and engineer

Nature is imagination itself.

William Blake, English poet

Oops, I just nearly tipped my chair over — I really should stop rocking
on it. This chair has four wooden legs. Round felt pads cover their base, so
they won't squeak. The chair's seat is made of polished wood, polished
both by the manufacturer and by my behind, as I write hour after hour.
The chair's backrest is a sheet of reeds woven around two wooden rods on
each of the chair's sides. All these parts make up the chair's *structure*. And
what does a chair do, what is its *function?* Easy, you will say. A chair is
made for sitting.

But what if I choose to put it to a different use? If the light over my
head flickers and goes out, I might stand on it to change the bulb. When a
cockroach scurries across the floor, I might squash it with a chair's leg. If
my laundry dryer breaks, I might dry clothes on the chair's backrest. But,

you might say, all these are not the chair's *real* function. Its real function is what its maker meant it to be, to be sat on.

Here is a very different example. A tiny finch on the Galápagos archipelago uses little dead twigs as tools. In its beak, these twigs get a new life. Twig in beak, the finch pokes around in tree cavities, hollows dug by burrowing insects that inhabit the tree. Stirred up by a probing twig, the insects scurry out of their shelter. Moments later, they die in the finch's beak.

What is the function of these twigs? To be sure, in their more glorious days they were part of a tree. They bore the leaves that allowed the tree to catch sunlight. Through their veins flowed valuable nutrients. And they carried flowers that helped the tree produce offspring. But that was all in the past. Eventually, these twigs died and fell to the ground, abandoned and useless. In a finch's beak, they gain, briefly perhaps, a new, quite different purpose.

What is the *real* function of these twigs? Does the finch have a role in determining it? Do we need a maker to breathe purpose into them? The examples of these chapters suggest an answer. They stand as metaphors for many questions we have about ourselves, about our place in the world, questions that humans have posed since time immemorial, questions that arise about everything from twigs to whole organisms, from bacteria to humans, and ultimately about the mind and the world itself. And the answer I propose — a place to stand — is that function and purpose have little to do with a maker. Once again, this is a perspective in which the living has enormous power.

The Open Purpose of Enzymes

I will begin with another look at enzymes, which are the molecules that facilitate the chemical reactions on which life depends. With such reactions, organisms change food or light energy into useful forms of energy — much as we use coal or sunlight to generate heat or electricity. These reactions also fabricate the building blocks, the molecular bricks a cell needs to grow.

Enzymes facilitate chemical reactions similar to how receptors and their messages communicate. Take a reaction joining two molecules, A and B, to create a third molecule, A-B.[1] An enzyme joining A and B contains surface patches that match A and B. These surface patches are near each other. A and B, when encountering the right parts of the enzyme, attach to the enzyme because their surfaces match that of the enzyme. And when they attach, the enzyme changes shape, much like a receptor reading a message. This shape change brings A and B close by. It may snap A and B together like two building blocks of a construction kit. The newly created molecule A-B then dissociates from the enzyme, leaving the enzyme free to join another pair of molecules A and B.[2]

Not all enzymes join two molecules. Some break a molecule into two pieces. Others rearrange the parts of A and B to form C and D. But the principle is the same. Molecules read one another's shape; they "understand" some shapes and interpret them by changing their own shape. This change results in the creation of new molecules.

What enzymes do seems clear on the surface: they facilitate chemical reactions. But many subtleties lurk underneath. Here is one. The molecules A and B whose reaction an enzyme promotes are usually much smaller than the enzyme itself. When attached to it, they cover only a fraction of its surface. The remainder of the surface is vacant. What if among the billions of small molecules in a cell, one — different from A and B — can attach to this free surface area, and what if the enzyme reacts with a shape change? Shape changes in proteins can drastically remodel the protein's surface. They may remove previously adjacent amino acids from one another's vicinity, and they may bring other amino acids close together. As a result, the molecules A and B may now attach far from each other to the enzyme, or the enzyme might not recognize them at all.

Small molecules that change protein surfaces in this way do exist. They are called allosteric effectors. In their absence an enzyme facilitates chemical reactions. In their presence it does not.

How do allosteric effectors fit into the simple picture, where certain proteins we call enzymes *really* promote chemical reactions? After all, when such an allosteric effector binds to a protein, the protein no longer facilitates a chemical reaction. One could still insist that enzymes really are proteins that facilitate chemical reactions, unless an allosteric effector prevents them from doing so. But consider that the allosteric effectors I just mentioned are only one of many kinds. Other such effectors do precisely the opposite. To be sure, they also bind to the surface of a protein. And in response, the protein changes shape. Until the shape change, A or B cannot attach to the surface, or they attach far apart from each other. After the shape change, they can attach near each other. When they do, the protein changes shape a second time and creates A-B. Such a protein promotes chemical reactions only when an allosteric effector is attached. To further complicate matters, a second allosteric effector may attach to the same protein, and influence what it does; or a third one; and so on. We know of as many as nine different allosteric effectors that can attach to the same protein. They may determine what kinds of small molecules A and B can attach, how fast or how firmly they attach; or they may even allow a protein to attach to other proteins, creating new protein shapes.

Pinning down the "purpose" of a protein that can communicate with as many as nine other molecules becomes a messy affair. For one thing, what such a protein does is not up to the protein itself. Rather, it emerges from interactions or relations that the protein enters with molecules in the world around it. And if this holds even for proteins — which are after all "just" molecules — how much more would it hold for other objects, objects we are more familiar with?

Thus far, I may have unwittingly suggested that certain enzymes can *at best* facilitate chemical reactions, that they cannot do something else altogether. Nothing could be further from the truth.

Consider, for example, eyes, such as those of a squid or a human or the thousand facets of an insect's compound eye. Eyes — like cameras — have lenses in them, which change the direction in which light travels, because the lens material is denser than the surrounding tissue and because light changes its direction when traversing materials with different densities. As a result, lenses can focus light. In our eyes, they focus light on the retina.

Camera lenses consist of glass or plastic. Eye lenses — get ready for this — consist of enzymes, the very proteins that facilitate chemical reactions elsewhere. This holds for organisms as different as humans and squids. In lenses, however, these enzymes do not facilitate chemical reactions. They occur only at extremely high concentrations, many millions of enzyme molecules in a small space. At these concentrations, an enzyme solution becomes a completely transparent jelly, much denser than its surroundings. Thus, enzyme molecules that catalyze chemical reactions in one tissue confer transparency and high density in another, enabling the lens to focus light.[3]

Moving Support Molecules

Enzymes are by no means unusually versatile molecules. They are not the exception to some rule. To see this, let us examine completely different kinds of proteins. These are cytoskeletal proteins, which give shape to living organisms.

Cells are to the molecules of life what houses are to humans. They provide space and shelter. Like houses, cells must not collapse. What prevents them from collapsing are the basic units of the cell's support system—its "bricks," or cytoskeletal proteins. They assemble in a rigidly regular pattern to form a support scaffold for the cell. Unlike bricks, they form not walls but long and thin molecular pillars, stiff rods somewhat like tent poles, but with two important differences. First, each cell contains innumerable rods, an incredibly complicated network, many more rods than poles in a tent. Second, their arrangement is much less regular than that of tent poles. They form a fine filigree, crisscrossing the cell in every conceivable direction. This scaffold gives the cell its shape and integrity.[4]

Again, the function of cytoskeletal proteins seems clear enough. They give shape to cells. But can they be put to different use, like their counterparts, the walls of human homes, which serve many purposes?[5] Adorned with paintings or tapestries, walls may decorate your living space. Mounted with bookshelves or cabinets, they may provide storage. They also hide unsightly or dangerous items, such as plumbing or electric wires that your dog might chew on.

To use walls to support, decorate, and hide is so obvious that we do not see the choices behind these uses. But think how much money you could save in electrician's bills if your house's wiring was readily visible and installed outside the walls. Alternative choices also exist for plumbing: an acquaintance who owns a two-story house chose to run a sewer pipe from the upstairs bathroom through his downstairs dinner table.

A cell's support rods are no less versatile than a house's walls. For instance, a cell constantly needs to ferry molecules around. It may produce molecular building blocks for its growth at one end, but it may need to use them at the other end; it may produce *export* goods, such as hormones or nutrients, that must get to the port of departure; it may need to *import* yet other molecules and ferry them to their proper destination in the cell.

Each cell of our bodies, during every moment of its life, transports innumerable molecules with different destinations. This task is similar in scale to that of a postal service handling millions of packages, a phenomenal logistical challenge.

How do cells transport their shipping goods? First they package them by enclosing them in tiny bubbles called vesicles. These vesicles are made of the same material as the membrane separating the cell from the outside world. To move packed membrane vesicles, cells use a highway network that reaches into every remote corner of the cell. It is an unusual network, the same cytoskeletal network of rods that also gives the cell support and shape and that reaches into every part of the cell. Cells move their vesicles along this network. A specialized protein called a molecular motor latches onto a vesicle and onto a support rod. It then glides along the rod and delivers the attached vesicle to its destination. Inside a cell, thousands of these vesicles are on the move at any one time, a giant symphony of elevators moving in all directions, gliding silently along the support scaffold that crisscrosses the cell. Thus, the same support rods that give shape to the cell can also serve as a transportation network.

In addition, these support rods can serve yet another function, apparently the exact opposite of giving support. They allow cells to move from one place to another.

Have you ever observed an amoeba under the microscope? Amoebae are amorphous, shape-shifting lumps of protoplasm that crawl by extending stubby protrusions at one end and retracting them at the other. What allows them to change shape constantly is the very network of support rods we are talking about. At the front end, an amoeba continually adds onto this network. Its millions of rods are growing ceaselessly through addition of individual protein bricks. At the rear end, these rods are disassembled, and their parts recycled. The extension of rods in front and their demolition in the rear create an amoeba's motion. That is, amoebas crawl through a constant remodeling of their support scaffold. It is as if you moved a house by constantly demolishing the back and recycling its materials to build additions to the front.

To summarize, cytoskeletal proteins play many roles in a cell. They confer stability; they form the transportation highways of the cell; they allow cells to divide; and they help cells move.[6]

Open Molecular Conversations

All these examples have the same theme. One and the same protein can serve vastly different purposes (like chairs, remember?). And the structure of any one protein influences but does not predestine the roles it can serve. The same holds not only for proteins but also for other molecules or for more complex mixtures of molecules.

Water-living creatures such as oysters and clams, for instance, are protected by hard, hinged shells that the animal can close instantly when in danger. To build a shell, these creatures extract building materials from the surrounding water. Their main building material is the calcium carbonate dissolved in water. But using a water-soluble building material for shell construction harbors a danger: water poor in calcium carbonate can readily dissolve a shell by leaching calcium carbonate from it. Consequently, the shell will grow thinner and eventually disappear. To prevent this death by dissolution, some shelled creatures protect themselves with a layer of waterproofing. This layer is made of a material called conchiolin. Conchiolin, however, can also protect against an altogether different threat.

Given how hard it is to pry open a living oyster, one might think that oysters would have few natural enemies. Yet a range of animals specializes in cracking their armor. Some fishes and crabs do so with little subtlety. They simply crush oyster shells in their jaws or pincers. Others — snails especially — doggedly drill away at a shell until they have bored a tiny access hole to the nutritious innards. But some shelled animals protect themselves from such assault.

Their protective material of choice is again conchiolin. Conchiolin is not only waterproof but hard enough to prevent crushing or drilling. At its evolutionary origin, conchiolin was probably used for only one of these two purposes, although we may never know which one.[7] More important for us, however, is that it can be put to two different uses, to protect against water and against predators.

The creativity of organisms in self-defense is a rich source for other, similar examples, including the defenses of some tropical vines against insects.[8] Beetles, ants, and other insects that eat these vines relentlessly haunt them. The most important and vulnerable parts are the flowers, because they are essential for reproduction. These tropical vines defend their flowers by exuding toxic or bad-tasting resins that deter many insects.

These resins can also serve another purpose. Some ants that live in large colonies actively collect these resins because they are waterproof, supple, and soft. Malleable yet waterproof building materials are a valuable commodity for these ants, because they can protect against the constant flooding that comes with the daily downpours in a tropical rain forest.

Proteins, conchiolin, and resins involve large and complex molecules. Such molecules consist of smaller building blocks (molecules such as amino acids in the case of proteins), which consist of even smaller molecules and atoms. We like to think of small molecules or atoms as "simpler" than larger molecules. Because of this simplicity, do these smaller parts have a more limited potential to do different things? Does their structure determine what they can do? Perhaps open purposes arise with complexity?

Consider this: neither one chemist nor an entire army of chemists would be able to list all interactions a molecule could have, not even for a molecule as small as water, with its two atoms of hydrogen and one of oxygen. Even if they focused on interactions between water and only one other kind of molecule (for example, the proteins inside one organism), the possible interactions would be innumerable, as most protein chemists could tell you.[9] And no smaller is the number of possible interactions for other molecules — however simple — whether inside or outside the living.

But perhaps we have not studied molecules long enough. Eventually, perhaps in a hundred or in ten thousand years, will we be able to list all possible interactions any molecule can be a part of? Consider that there will never be such a list. One way to see this is by considering the accomplishments of synthetic chemists. For centuries, chemists have been creating new molecules. My computer's keyboard, its liquid-crystal display, the ink and paper of your book, the coolant in a refrigerator, high temperature superconductors, composite materials in airplanes — all of these human creations contain molecules that have changed our world. Many of these molecules exist nowhere else in nature. And nothing indicates that human synthetic chemists are slowing down. If anything, they are speeding up, creating more novel molecules every year. If the number of interaction partners is growing constantly and ever faster, how will we ever come up with a complete list of possible interactions for any one molecule?

In addition, life on earth is vastly more prolific at synthesizing chemicals than any chemist. Earth may harbor more than ten million species, of which we know only a small fraction. Every new species we discover may produce molecules unknown to us that allow it to feed on exotic nutrients, communicate with other organisms, or fend off aggressors. And novel molecules are undoubtedly being created somewhere, by some organism, while you read these lines. Life's creativity vastly outpaces our modest abilities to discover its magnitude.[10]

Creation is not only prodigal but ongoing. This alone would suffice to demolish any fantasy of reducing the world to Lego-like simplicity, to a construction kit with entirely predictable parts. The dream of some scientists — the nightmare of others — of reducing creation to predictable interactions of simple mechanical building blocks is a pipe dream.

Two additional factors conspire to lay waste to this dream: time and space. As a whimsical example, assume there was a protein that occurs only in polar bears. The protein's shape could promote a chemical reaction between two substances, A and B. However, polar bears contain neither A nor B. Both occur, say, only in some tiny weevil in the Amazon rain forest. They will never get close enough to a polar bear to encounter this enzyme. And this enzyme, like everything else — chairs, polar bears, weevils, and molecules included — does not exist forever. How could we identify all of its possible interactions in an immense world of things separated by vast swaths of space and time?

Beyond Molecules

Eyes are inexhaustible sources of insight in many ways. Their example shows that living matter follows the same principles as the dead matter we have just examined. Eyes like ours have an iris, which is an elastic, multi-colored, contractile ring of muscles that changes the pupil's diameter and so adjusts the amount of light illuminating the retina.

Irises also play other roles, one of which involves repairing eye damage. Although humans cannot regenerate damaged eyes, many other organisms can. Some can even replace entire eye parts, such as the lens.

Recall that lenses form in an interaction between two groups of cells, the neural tube and ectoderm, which exist only in the embryo. The neural tube later becomes the spinal cord, the brain, and the retina. The ectoderm transforms into other body and eye parts, including the cornea, the transparent outer shell that protects our eyes.[11] Embryonic ectoderm and neural tube do not exist in the adult organism. How can lost lenses be rebuilt if these tissues no longer exist?

Rebuilding lenses requires a process completely different from the one that creates lenses. In organisms such as salamanders the iris is critical in this process. Here, lens replacement begins with a small group of iris cells. These cells first lose their color. They then begin to divide, forming a larger and larger ball-shaped outgrowth of the iris, reminiscent of a tumor. As this ball grows, its cells begin to produce the proteins responsible for lens density and transparency, the very proteins I described earlier. Eventually, this ball detaches from the iris and replaces the lens.[12]

The Creation of Function (and Meaning)

As a group all these examples point to some general principles. First, the kinds of interactions in which organisms and their parts are involved and the kinds of functions they can assume are limitless. Only our imagination to ask the right questions about them is limited. Second, whether the part is a small molecule or a large organ, its functions are not properties of the part itself. They emerge from the relations or interactions this part has with the world around it. The two roles of an iris I have just discussed illustrate how, when interacting with light, irises regulate flow, but when interacting with damaged lens tissue, they serve to repair. A cytoskeletal protein's interaction with the cell's surrounding membrane supports the cell; the same protein's interaction with vesicles serves in transportation. We can only understand what parts of a living organism do in relation to other parts. Their purpose is not written into their structure.

Philosophers and naturalists have thought and argued for centuries about the purpose of living organisms, their parts, and whether purpose might be carved into these parts themselves. Why? Because if so, this points to an intelligent designer — however distant — who has endowed life with sufficient ingenuity and foresight to persist for billions of years and perhaps even with a final goal and purpose, however incomprehensible to us.[13]

Faced with the marvelous complexity of life, the idea of intelligent design is indeed seductive. Hearts pump blood more faithfully and tirelessly than carefully engineered pumps. Leaves collect sunlight with astonishing efficiency. Eyes are incredibly versatile cameras. Nevertheless, most biologists now reject the idea of intelligent design, for various reasons that others have written about at length.[14]

The examples I have presented here also speak against predetermined design and purpose. All are about parts of organisms that can serve different purposes and carry out different functions depending on the conversations in which they participate. Not only that, but an organism's parts change their uses over time. The three tiny bones that transduce sound in your middle ear once served as jawbones in primitive fish. The same bony limbs that dragged primitive amphibians across land allow birds to soar sky-high. The enzymes that have been promoting chemical reactions since time immemorial later became part of eyes that see. Many of the thousands of proteins inside any living cell may have changed their functions innumerable times during their billion-year-long journey inside the living.

There is much to admire in the design of living organisms. But to admire only design is to take a narrow view. It is the admiration we reserve for a beautiful crystal, a well-engineered machine, or a fixed specimen under the microscope. Only a broader perspective reveals the prodigal creativity inside and around organisms and the ever-changing purposes bestowed on their components. From this perspective, we see an intricately constructed epic. Each character in this epic assumes ever-changing, complex roles. And in contrast to many other epics, *this* epic may be unending.

Ceaseless and prodigal creation excludes predestined purpose. But the absence of such purpose raises thorny questions. One is where function originates, if it is not predestined by some distant designer. Do function and purpose—however fleeting—apply to colliding molecules? That might be a stretch. However, we can be sure that function and purpose are at least as old as life itself. A finch lending new purpose to fallen twigs or any animal using tools might serve as examples. What I have in mind, however, goes far beyond these commonplaces. Any enzyme that produces energy from a food molecule has acquired purpose through the organism using the enzyme. So have cytoskeletal proteins that maintain cell structure, the receptors that read molecular words, and the million other parts of organisms I have not talked about.

The same holds for whole organisms and the relationships they have with others, whether as predators and prey, host and parasite, or in some mutually beneficial way. The closer we get to organisms like us, the more obvious this feature of endowing new meaning and purposes becomes. Closest to home is of course human creativity — manifest in art, literature, science, and technology. Much of this creativity gives the old new purposes, combines the existing to make the new, and thus creates new relationships, over and over, until ancestral purposes become completely eroded.[15] Consider that human creativity and life's creativity, creativity that gives new purposes to everything from molecules to organisms, are based on the same principles. They differ only in degree, and many of these differences crumble on closer inspection.[16] For instance, some might argue that only human innovations have truly transformed the world. Not true. Other organisms change the face of the earth much more radically than we ever did. Think of organisms capable of photosynthesis — before them there was no oxygen to breathe — and the innumerable microscopic organisms that till and prepare the soil we use to grow our foods.[17]

A second question emerging from the absence of predestined purpose is this. What *is* function? What *is*, say, the function of an organism's part?[18] If I have rather loosely referred to functions and purposes, it is because this question in its most general form is unanswerable. Some might say the answer lies in why the part originated and why it is still in existence.[19] But this approach would not get you very far. It fails if one part plays a role in more than one aspect of an organism's life. It also fails if the part's roles change over time, as is usually the case. Every other attempt to arrive at a precise definition of function has its own Achilles' heel.[20] But more important for us, an answer to the question "What is function?" requires a vantage point that eliminates the novel, surprising, and serendipitous from an ever-changing world. I choose not to stand in this place, because it is much more important to appreciate the ever-changing and innumerable relationships among organisms and their parts.[21]

The Price of Openness

Is there a price for the absence of confining design we see from molecules to organisms and for the astounding creativity sprouting everywhere? I am afraid there is: life does not always turn out as a benevolent designer might wish. We have encountered this issue before. Remember parasitoid insects, which lay their eggs inside other organisms, where they live and feed on their hosts? These larvae often keep the host alive, barely alive, but enough to feed itself and thus its inhabitants, until they kill the host. From the perspective of a parasitoid, this is a marvelously ingenious way of keeping the food supply fresh and even growing. From the host's and from our perspective, it is a horrendously exploitative lifestyle, a cruel way of killing for a living. Yet as many as two-thirds of all organisms on this planet are parasites. Most do not live as eccentric a life as parasitoids. But many have no less ingenious ways of taking advantage of their hosts. And when examined closely, free-living organisms (the opposite of parasites) generally take advantage of others as well.[22]

A parasite's lifestyle, like anybody else's, serves the parasite itself. Thus, one could argue, if there was a designer, it has at least designed lifestyles with somebody in mind, although at somebody else's expense. But often not even that is the case. Were you ever stung by a bee? The sting hurt you, but it hurt the bee even more. A bee's stinger is barbed; when retracting it, a bee rips its guts out. Barbed stingers have no benefit to anybody, because a stinger without a barb or with a retractable barb would serve its purpose just as well.

As a last example, consider tumors and the horrid deaths they cause. Among many gruesome ways of dying, a slow and pain-racked death from cancer is perhaps the most undignified. And yet, no organism may benefit if you or I have a tumor.[23] This stands in contrast to infectious diseases, which may aid the bacteria and viruses that cause them. A tumor's main characteristic is that cells divide rapidly and uncontrollably. Ensuring such rapid cell division is not easy. For example, like organs, solid tumors need a nutrient supply. And the bigger they grow, the more nutrients they need. To solve this problem, some tumors build a sophisticated network of blood vessels to feed themselves. This network is part of their astonishingly efficient and ferocious way of plundering our bodies for nutrients. In other words, there is method to the havoc tumors wreak. You could view tumors as parasites of our body, because in furthering their own growth, they extract nutrients from the body. However, they are unusual parasites because they arise from within a body. And when they kill that body, the tumors go down with it and leave no offspring.[24]

In sum, a sophisticated way of living that we marvel at is often an abomination for those at its receiving end. Often such sophistication even leads to something utterly shortsighted, pointless, and destructive.[25]

Barbed stings and a death in agony are the price we pay for being free from eternal confinement to a predestined purpose. It is the price of freedom. This choice of perspective has two consequences. The first is to accept our ignorance — blissful or dreadful — about the world's future. The second is perhaps more disturbing: to choose an unknown future also means to admit the possibility — however remote — of a final purpose and goal to life. However, to then lock oneself — by choice — into this last perspective is to walk in a circle and abandon one's power. Here surfaces again the Gordian knot of the paradox. Short of the sword of choice, there is no way to unravel it.

CHAPTER 8

Choice and the Natural Sciences

No new principle will declare itself from

beneath a heap of facts.

Sir Peter Medawar

Among all the conversations from the microscopic to the planetary realm that build the world, one is special to humans in our time, and it has shaped our world unlike any other. It is the conversation we call natural science. Antibiotics, electric lighting, computers, telecommunications, air transportation, and much more are its products.

Why is science so powerful? What distinguishes it from other conversations? These questions occupy this and the next chapter, where I argue that science creates and requires especially powerful and complex choices.[1]

The Primacy of Explanation

I am concerned here with science as a conversation with nature, whether human or nonhuman. In this conversation a mind poses questions to nature and receives answers. These answers might consist of an elementary particle smashing into a detector, thus revealing its existence; they might consist of a plant turning its leaves toward a window; or they might consist of a questionnaire about the importance of science, filled out by a hundred interviewees. This conversation aims to organize the world. That is, it aims to *explain* the world's features and *predict* its future.

Among these aims I focus on explanation here, for two reasons. First, explanation — right or wrong — comes before prediction. Newton's law of gravitation not only explains why earth revolves around the sun but permits the prediction that it will take 365 days to do so. To predict, one must have explained first, however crudely.[2] Second, sometimes we cannot predict, even though we can explain. Take the notoriously unpredictable weather. We can explain it from simple concepts such as heat, humidity, and solar energy. But we cannot reliably predict it. And recall that even simple objects like three planets spinning around one another can defy prediction, although their force of attraction is perfectly explicable through Newton's law. For these two reasons, I consider explanation — the identification of causes — a more profound aim of science.[3]

The Endless Maze of Explanation

Scientific conversations start from a description of the world, from basic facts, from the "what" and "where" of things. Their goal is to find the "why," the causes, the explanations behind these descriptions. At first sight, description and explanation are thus different animals, one being the start and the other the end of a conversation. But examined carefully, description and explanation are closely related: any description already implies an explanation, an interpretation of the world. And any explanation is a description, as some examples will show.

Why is this important? Because it means that the departure point of our conversations — the facts that we do not question any further — are not clear-cut. They can become matters of choice. It also means that a scientific conversation heaps explanations upon explanations upon explanations (or descriptions upon descriptions upon descriptions). It creates an unending story about the world. And in creating this story it wanders deeper and deeper into the maze it is building. Any example of a scientific conversation could serve to explain — or to describe? — this relation of explanation and description. So let us start somewhere, anywhere.

Until World War I, most wounded soldiers died not from their wounds but from deadly infections, such as gangrene. To prevent such deaths has always been a prime goal of battlefield medicine. Imagine you are a physician dedicated to this goal. To learn about the enemy, you study bacteria that infect wounds. You culture them in the laboratory in petri dishes, small round dishes containing a nutritious jelly. The nutrients, however, also attract fungi whose invisible spores coast through the air in the millions. Unless you are careful to keep your bacterial cultures clean, they will rapidly become moldy. One day you discover, by accident, that a mold has infected one of your cultures. To your astonishment it has also killed the bacteria on the dish, and you ask why.

Something like this beginning of a scientific conversation actually took place. The year was 1929, the scientist was Alexander Fleming, and the fungus was *Penicillium notatum*. The ensuing conversation culminated in one of the greatest breakthroughs in medicine.[4]

Why did Fleming's bacteria die? Did the fungus eat them, or did something the fungus produces—a molecule perhaps—kill them? You could pose this question in many ways. You could culture the mold itself in a liquid medium and then pass the medium through a filter so fine that the fungal cells cannot pass. Is the resulting filtrate still able to kill the bacteria? If so, then intact fungal cells are not required for the bacteria's death. Alternatively, you could kill the mold by boiling it. Do its boiled remnants still kill bacteria?

There are many other ways to ask this question, and they all lead to the same explanation: the bacteria die from a molecule the fungus produces. This molecule is called *penicillin*. This explanation—it can also be viewed as a description—immediately invites the next question. Why does penicillin kill bacteria? Multiple possibilities exist. Penicillin might destroy a protein essential to bacterial life, prevent copying of the cell's genetic material, inhibit cell growth, and so on. The answer, it turns out, is close to the last possibility: penicillin prevents bacteria from growing and remodeling their cell wall, which is their life-saving barrier against a hostile world.

You can also view this explanation as a description: penicillin is a molecule that prevents bacteria from adding to their cell wall. How? A bacterium's cell wall consists of long stringy molecules resembling a tangle of spaghetti. Short molecules that connect or cross-link these strings give this tangle the stability of a wall—or at least of a strong mesh. No cross-links, no cell wall. Penicillin stops the cell from cross-linking these molecules. How so? By blocking a protein, an enzyme called transpeptidase, from fostering the necessary chemical reaction. How? Penicillin binds to the surface of transpeptidase and thus prevents the molecules to be cross-linked from binding there.

These are only six of countless levels of explanation, answers to "why" or "how" *Penicillium notatum* kills bacteria. Each level can lead to many further questions whose answers lead to further questions and so on.

Taken together, these levels of explanation can also be viewed as an elaborate description: *Penicillium notatum* produces a molecule that binds to transpeptidase, which cross-links molecules that form the cell wall, which is necessary for bacterial life. Conversely, consider that you can view most descriptions as explanations. Descriptions are the other side of explanation. If you practice this with any description/explanation, you will find that it usually works.

The apparent exceptions are such descriptions as "A ball is round," "This knife has a jagged edge," and "This surface is cold." Are these not just basic "facts of life"? Are they not indisputable descriptions? Most people thought so, until cognitive psychology developed as a discipline and taught us otherwise.

For example, consider a simple statement about color: the mold *Penicillium punctatus* is yellow.[5] Had Fleming—who made this statement—routinely observed his mold next to a bright red object, the mold would have appeared black. Had he routinely observed it against a black background, it would have appeared white. And had he been interested in the nature of light, he might have passed the yellow light through a prism. Prisms not only split white light into a rainbow of different wavelengths. They also split light of any apparent color into its component wavelengths.[6] The prism might split the light from some yellow mold cultures into one blue and one red light beam and the light from others into some mix of blue, green, and red. In general, infinitely many mixtures of pure colors appear yellow to us. Our perception of light as yellow is an interpretation of our eyes and brain.

The same holds for all of our senses, although it is best known for vision.[7] An enormous amount of interpretation, beginning with our eyes, ears, and noses, takes place before we see any color, hear any sound, or sense any smell. Our sensations are by no means basic facts, primordial descriptions of the world. Our immediate experience of light, sound, and touch thus does not provide an exit to a sprawling labyrinth of explanations/descriptions.[8]

The Creation of Choice

Our sensory interpretations of the world must surely differ from those we call scientific explanations. For my purpose, they differ in only one key point: we may have little or no choice in making them.[9] Consider that aside from this difference, the explanations emerging from a scientific conversation share much with our sensory experience.[10] And consider that the same holds for nonhuman interpretations of the world. These are the interpretations I earlier called living worldviews, tacit and embodied explanations.[11] Recall the chick that tacitly explains a shadow hovering overhead—by ducking and hiding. To the chick, a bird of prey *causes* the shadow. Or take the parasitic plant that chemically senses the presence of a tree. This presence means "food" to the plant; otherwise it would not invade the tree with its roots.[12]

Whether it is a bacterium swimming for food, a parasitoid depositing its eggs into a living caterpillar, a predator stalking its prey, or a cell in your body blowing itself up, these actions reflect interpretations or explanations of the world—whether chosen or not. In all of them, an object or event comes to stand for something else, something that comes under a variety of names: explanation, cause, interpretation, or simply meaning.[13]

Some human and nonhuman interpretations cannot be modified by choice. In others, choices exist and are obvious. Choices can also sometimes disappear, such as when chosen interpretations become automatic. If you have ever studied a foreign language, a musical instrument, or learned to drive a car, you know what I am talking about.[14] But more instructive is the opposite, the creation of choice. Such creation can have far-reaching consequences. It demonstrates the power of choice, one of my central themes.

Scientific conversations are ideal to illustrate this power, because the boundary between chosen and other interpretations is most fluid here and because science is replete with explanations, unquestioned descriptions, and even "facts" that are either never challenged or have become dogmas. The most invisible among them are the most dangerous, because they impede choice.

Every once in a while, some human being creates a powerful choice of perspective that did not exist before, a choice that nobody else could see before. At their best, such choices can revolutionize how we see the world and form the seeds of new technologies. Here are three concrete examples.

A Timely Choice

Perhaps the most dramatic examples of how human choice changes the world come from physics. Because the claims of physics are to the world as a whole, not only to the history of earth or to the history of life, its perspectives apply equally to the submicroscopic world of atoms and to galaxies; to the present, to the past, and to the future. Perhaps because of the all-encompassing generality of physics, many historians have studied how the perspectives of physicists change over time. And every major development in physics could illustrate how powerful individual choices can be: Copernicus's choice of earth as one of our sun's planets;[15] Newton's choice of viewing distant bodies as gravitating toward one another; Maxwell's choice unifying electricity and magnetism into electromagnetism.

An especially striking example involves Albert Einstein, who chose a perspective that none of the physicists at his time had chosen — even though the observations he relied on were public knowledge.[16] In fact, in Einstein's time physics as a whole had reached a dead end. Many physicists thought that the major problems in physics had been solved. What remained to be done was to mop up the "details," many of which were observations the existing theories had not explained. Most physicists believed that minor fixes of these theories would eventually take care of such details.

A particularly stubborn detail had to do with the speed of light. Think of light as a wave carried by a medium, as water waves are carried by water. Physicists did not know what this medium was. But many believed that such a medium must exist, and so they gave it a name. They called it the ether.

If you sit in a boat on the surface of a large lake, and somebody drops a rock some distance away, the rock will make waves. The wave crests lap against your boat at some speed. This speed will depend on whether your boat moves in the water. If the boat moves away from the rock, the waves will come toward you more slowly than if the boat stands still. If the boat moves exactly at the speed of the waves, the waves will appear to stand still. If, on the other hand, the boat moved toward the rock, the waves would appear to come at you faster.

Physicists applied the same line of thinking to light and the ether. With one important difference: unlike water, we cannot see or feel the ether. And thus we cannot easily measure our speed relative to the ether. We can measure the speed of light, however. In addition, whatever this ether is, there are good reasons to believe that earth moves relative to it, like a boat moving in water. To see this, keep in mind that earth rotates not only around itself but also around the sun. And it does so at the break-neck speed of several thousand kilometers an hour. Thus, unless the ether follows all of earth's motion with exquisite precision, the earth moves relative to the ether. And not only that, the direction of this movement changes continually, precisely because the earth revolves around the sun. This means that the speed of light coming at you from some immobile source on earth should change as the earth's position relative to the ether changes.

Physicists made great efforts to measure the variation in the speed of light caused by the earth's motion.[17] However, they found no such variation. Regardless in which direction a light beam moves relative to the ether, it moves at the same speed. This was a stubborn "detail" hard to explain with any of the existing perspectives. In order to do justice to this observation, physicists needed to abandon dearly held principles.

Some physicists argued for minor fixes to the existing perspectives. One of these fixes was the idea of an ether drag. According to this idea, earth drags the ether behind it as it moves, much as a moving boat would drag water behind it. Thus, if you were on or near the earth, earth would appear still in relation to the ether. Others argued that the ether was a flawed idea and should be abandoned.

Albert Einstein made the most radical choice of all. He not only abandoned the idea of the ether. He also abandoned a time-honored principle, the principle of absolute time. Essentially, he proposed that if you move relative to a light source, then a clock moving with you runs at different speed than a clock at the light source. In this perspective, a constant speed of light no longer poses a problem. If your boat moves away from where the rock was dropped, time elapses more slowly, exactly so much more slowly that the waves move toward you at the same speed—distance covered per unit of *time* — as if you were at rest. This choice was radical because it spoke not only against accepted wisdom but also against intuition: nothing in our daily experience tells us that moving clocks run differently.[18] And it has many profound consequences, such as that the lengths of moving bodies also change and that no body can move faster than the speed of light. Physicists later confirmed many consequences of Einstein's theory. His choice transformed how physicists look at the world.[19]

Land at Sea

Powerful choices have also shaped the face of other sciences. The following example from geology starts with a seemingly simple question: Why are there mountains? Humans stand in awe of them, their unimpeachable majesty and timeless magnificence, their merciless indifference. Our very insignificance in the face of their vastness might have made any questions about their origin — other than by divine creation — a sacrilege. But it did not. People questioned the origins of mountains since the earliest days of geology. And in the nineteenth century, this question absorbed many geologists — perhaps because mountains are literally hard to overlook, and perhaps also because they have immense economical importance as a source of minerals.

In the nineteenth century and far into the twentieth, geologists accepted the following answer to the origin of mountains.[20] The earth, in its early history, was a fireball of molten rock. It steadily lost heat to the frigid space around it until the thin solid skin on which we live was formed. Underneath this crust, however, the earth is still seething, and it continues to radiate heat into space. As a result, like any other cooling object, the earth is slowly shrinking. And its skin, our familiar crust, shrinks with it, like the skin of a slowly drying apple. Some four billion years after its origin, this shrinking apple has low-lying areas. These are the ocean floors. And it has elevated areas, the continents. In many places, ripples are forming while the apple shrinks: the Himalayas, the Alps, and the Andes are but wrinkles on the skin of a drying apple. Mountains are thus what remains after the rest of the crust has sunk closer to the earth's center.

Here is an important consequence of this perspective. If the surface of the earth forms as the crust sinks vertically — and only vertically — all surface features should remain in the same place. That is, all the continents should be where they were, say, two billion years ago; so should the ocean floors; and so should the mountain ranges.

Although widely accepted at its time, this "shrinking apple" theory—geologists call it the thermal contraction theory—had problems, like any other theory. At best, it could not explain many observations. At worst, some observations seemed to contradict it outright.

One among several problems with this perspective is that very similar plant and animal species inhabit parts of continents separated by vast swaths of ocean. For instance, the island of Madagascar hosts many species of lemurs that are also common in India. Australian marsupials are almost identical to their South American counterparts. (They even share some of the same parasites!) Yet Madagascar is much closer to Africa than to India, and Australia much closer to Asia than to South America. How is it that landmasses separated by thousands of miles of ocean harbor similar species? And that these species are sometimes more similar between distant than between close landmasses?

Many species, including numerous ferns, reptiles, and earthworms, fit this strange pattern. The case of earthworms is especially striking. Earthworms can not swim or fly, nor do they have airborne seeds, spores, or water-living larvae that could float from coast to coast. How did these species cross thousands of nautical miles to get from one landmass to another?

A second problem stems from the folding of rock layers in mountain ranges. If mountain ranges are what is left after the crust has sunk, different types of rock should be layered in an orderly way. Like the layered sheets of pasta, sauce, and cheese in lasagna, individual sheets of rock—different by chemical composition or age—could be warped, but their order should be preserved. For instance, newer rock layers should be stacked on top of older ones. But geologists often observed otherwise. In many mountain ranges rock layers are folded like an accordion, as if an invisible hand had squashed the lasagna, very slowly but with immense force, from both sides. In addition the folds of this accordion lie sideways in many places, as if they had fallen over. In such places rock layers of different age are intermixed, with not just younger layers stacked atop older layers but older layers piled atop younger layers. Some of these folds extend over hundreds of kilometers. It is easy to see that a gargantuan force must have produced them. It is much harder to understand how mere sinking of the crust could have.

These are just two among many problems the thermal contraction theory faced. Although many geologists thought that eventually the theory might solve all of the problems, some thought otherwise. Notable among them was the Austrian geologist Alfred Wegener. He chose to reject the theory outright. In his book *The Origin of Continents and Oceans*, Wegener chose a completely different perspective.[21] He argued that we have to imagine the continents as plates of rock on a liquid surface, the molten rock of the earth's core. A little bit like leaves on a puddle, these plates float on top of the infernally hot core. And just like leaves, the continents drift this way and that, erratically. Some might run into each other, others might drift apart. But they certainly do not stay in place.

This perspective makes most problems with the older theory disappear. Why do similar species occur on different continents? Because at one point in their past, these continents were much closer together. They may have been part of one continent that broke into several pieces. The theory also provides an alternative answer to my opening question: Where do mountains come from? As continents — immense masses of rock — collide with each other on their endless erratic journey, they exert enormous force against each other. Think of two crashing cars and the force that transforms them into heaps of twisted metal. The difference is that this crash occurs in imperceptibly slow motion, over many millions of years. In Wegener's view, the invisible force that creates mountains is the same force that bends and twists its rock layers into byzantine shapes. It is the force of continents pushing against each other.

This perspective was a radical choice of one human being. It was radical because it departed completely from accepted wisdom, a body of thought built on the work of hundreds before Wegener. Choices like this require the chooser to be prepared for a lot of opposition, for years at best, for a lifetime at worst. However, although a radical choice like this can destroy a career, the potential rewards are equally immense.

When Wegener proposed his perspective, complete rejection, outright hostility, and eventual oblivion seemed all but certain. In Europe, his home continent, Wegener gained few followers. And the situation was worse in the United States, where many of the most influential geologists at the time almost uniformly rejected Wegener's idea as wrong, impossible, even pernicious and unscientific. Almost as imperceptibly as the rise of a mountain, however, his perspective gained ground. After decades it began to win converts. And today Wegener's theory of drifting continents is accepted wisdom. The old thermal contraction theory is a thing of the past, another casualty on a battlefield of human worldviews.

Choosing Change

Another prominent example of powerful choice is a theory central to biology, Charles Darwin's theory of evolution.[22] As is often the case, long before Darwin published his theory in 1859, many of the facts — descriptions — essential to the theory were known. First, a great variety of fossil organisms was known to occur in rock layers of different ages. Some of these fossils look like no living thing we know today, demonstrating that organisms have changed radically over time. Second, anatomists who studied body plans of different organisms had found many similarities among superficially very different species. Especially striking were similarities in the arrangement of bones. Take our forearms and the wings of bats: it is no surprise that their shapes differ radically, because bats use their forearms to fly. But the arrangement of their bones is strikingly similar to ours and shows a one-to-one correspondence between human and bat bones. Or take the bones in our legs: they appear in radically transformed shapes but identical arrangement also in the legs of hoofed animals.

A third key observation came from plant and animal husbandry. People had successfully bred animals and plants for thousands of years. They selected desirable features — fast legs, large seeds, beautiful fur, or simply good nature. And they allowed only individuals with these features to produce offspring. Among the offspring, breeders would select only those individuals with the most desirable features. And so forth. Over many generations, people had thus generated dramatic changes in many organisms. Modern food crops — cereals, vegetables, and fruits — look nothing like their wild ancestors and produce vastly higher yields. Most dog breeds hardly resemble the wolves they descended from, and their size range alone — from Chihuahuas to Great Danes — is spectacular.

All this was known before Darwin's time: the deep similarities among the living, its changing forms, and the profound effects human breeding has on these forms. Darwin wove these and many other observations into his theory of evolution. His choice of perspective included a selection process similar to that of human breeders, except over vastly greater amounts of time. This process generates the diversity of organisms from one or few common ancestors. Darwin's choice turned the world of organisms inside out and radically transformed humans' place in it. It was again a risky choice, and a choice that met much opposition. But its impact was as profound as the risks of making it.[23]

Who Chooses the Questions?

The questions we ask steer the direction of scientific conversations. But what steers the questions we ask? Some social scientists maintain a radical perspective: the times we live in and the people around us — from parents to teachers to peers to society at large — determine our perspectives. Therefore, they also steer the questions we ask. And not only that, they determine how we read nature's answers and the questions we ask next. Society is all that matters to scientific conversations.

Many scientists disagree with this perspective, and for good reasons. Here is its blind spot: the world's answers to profound questions can be surprising, puzzling, and seemingly nonsensical. Most of us follow our beliefs and gloss over such puzzles. We explain them away as anomalies, as exceptions to some rule. But some individuals recognize surprising answers for what they are, golden opportunities. If Fleming had dismissed the dead bacteria on the infected petri dish as an "anomaly," penicillin might never have been discovered.

The examples of Einstein, Wegener, and Darwin illustrate that the history of science is full of powerful choices by individuals who often went against accepted wisdom.[24] Thus, even though the world around scientists may influence the questions they ask, scientists continually break new paths. The biochemist and Nobel Laureate Albert Szent-Györgi put it best when he defined discovery as seeing what everybody saw and thinking what nobody else thought.

Swimming bacteria and the question what makes them swim for food provide an analogy that speaks to this issue. To argue that only society matters to the direction of scientific conversation is akin to the perspective that bacteria swim toward a food source because other bacteria have not consumed the food at the source. Thus, arguably, other bacteria determine where a bacterium swims—a perfectly valid perspective, but just that, a perspective, and not the one and only truth. For its blind spot neglects the many other parts of the whole, down to protein and food molecules. And, as we saw, much space exists between small parts of a whole and the impositions of this whole. An individual and its choices can fill this space.

The Power of Choice

Countless theories other than those of Einstein, Wegener, and Darwin could illustrate the key role of choice in scientific conversations. They include the notion that the earth circles around the sun, that many diseases are caused by infection (an alien concept until Pasteur's time), that our bodies carry a substance that can be inherited, that lightning and electricity are closely related phenomena, and so on.

One might think that the choices behind such revolutionary insights are determined by certain incontrovertible facts. Anyone educated enough who grasps these facts could not help but choose the new perspective. But choices of perspective on the world are every bit as messy as difficult choices in other areas of life. Part of the reason is that the facts are innumerable. How do you choose the few that matter? And sometimes the available observations themselves—the "facts"—are disputable, especially when different scientists made conflicting observations.[25] It is thus perhaps less surprising that scientific choices are difficult and that only few individuals make choices that transform the world.

Second, how do you know that a radical choice is necessary to explain existing observations? You don't. All theories face observations they cannot explain. More often than not, existing theories can accommodate such observations with minor fixes. Examples include the notion that distant continents were previously connected by land bridges, which could rescue the thermal contraction theory, or that the earth drags the ether behind it while hurtling through space, which could explain the unchanging speed of light. A radical choice of perspective is meritorious only in hindsight, years or decades later.

In addition, when making a radical choice, you confront an army of detractors. The reasons are many, but they boil down to this: every powerful choice requires giving up something humans hold dear. The detractors are not willing to give this up. They may not even be able to conceive of a world without it. This something might be that the earth is the center of the universe, that continents do not move, that all organisms were created at the same time, or that time is absolute.

But why do we defend our interpretations so fiercely? The answer, as we saw earlier, may lie in our ancestors. They, from bacteria to mammals, literally lived and died by their interpretations. It is nearly miraculous that we are able to choose at all, that we are free to separate our worldview from our fate.[26] Sometimes, I should say, because many humans still stake their lives on their views. The countless armed conflicts fought over religious beliefs or political ideologies and the millions of humans who have perished in them testify to the enormity of the task, to see interpretations as what they are. Scientific interpretations of the world are no exceptions.

In summary, human and nonhuman interpretations, descriptions, and explanations of the world are very similar in many respects. They differ in the role of choices, difficult but possible for us, impossible for many others. Choices are key to driving scientific conversations forward. The most powerful and difficult choices cause scientific revolutions. Choices build the edifice of knowledge, beginning with a description and a question whose answer adds the next brick to the edifice. Although you could argue that this edifice is an endless maze with no way out, you could also say that choice itself provides the exit. It is the choice of calling one interpretation the ultimate one. It is the choice of stopping somewhere, of ignoring much of an endless cabinet of mirrors, of focusing on one of its million reflections.[27]

CHAPTER 9

The Limits to Knowledge

> The aim of science is not to open the door to
> infinite wisdom, but to set a limit
> to infinite error.
>
> *Bertolt Brecht*

Scientific conversations have two advantages over other conversations. Both come from two rules of how to pose questions to nature: first, scientific questions must encourage nature to answer "no"; second, it must be possible to ask nature the same question over and over again. Both sources of strength have the weaknesses that come with any strength. After having outlined how these weaknesses limit our ability to know, we will return to the thread of paradoxes that runs through our conversation.

The Downside of Taking "No" for an Answer

Early in the twentieth century, the philosopher Karl Popper pointed to a simple but devastating fact about scientific theories: we will never be certain that they are true.

Suppose Alexander Fleming had elevated the action of penicillin to a law of nature: "penicillin kills all bacteria." How could he be sure? Bacteria occupy every tiniest nook on earth, from environments familiar to us—

soil, air, water — to others almost as hostile as deep space, infernally hot deep sea vents that spout toxic gases and scalding water or rock layers hidden deep in the earth's crust. The bacterial species living in all these environments may number in the tens of millions. How could we learn whether penicillin can kill all of them? Only by exposing all of them to penicillin. But that is impossible.

The same holds for all theories, from physics to the social sciences: we cannot verify that they hold everywhere we expect them to hold. Certainty eludes us. We cannot even be sure that the sun will rise tomorrow even though it has been rising every day for the past four billion years. So what can we be sure of? Popper argued that we can be certain only when a theory or hypothesis is false.[1] It is false when we observe a violation, such as bacteria unfazed by penicillin. (Fleming found even in his laboratory that penicillin would not kill some bacteria.)

This simple insight guides the kind of questions we must ask of nature.[2] Consider a simple example. Your neighbor's bright red car is always parked outside a street-facing window in your house. You observe that any plant you place in the window turns its leaves toward the car. You have tried this a hundred times and it works without fail. "All plants turn to face red cars" is thus a great candidate for a law of nature. Could you be sure? Never, because the world is host to too many plants. But you can pose the question by painting the neighbor's car blue.

The plant you place in the window the next morning promptly turns its leaves toward the window. As a result, you know that your "law of nature" is false. And you are left with nothing, except your neighbor's fury about his newly blue car. Maybe some chemical the car exudes, its smell, makes leaves turn? (You ask this question by sealing the house or moving the car.) Is it the glass in the kitchen window? (You open the window or remove the glass.) Maybe it is the heat generated by the sun? (You blow cold air onto the plant.) None of these questions can be answered with a conclusive "yes," but each of them can receive a conclusive "no."

Scientific conversations move forward by eliminating the flawed. And this is why questions that allow a "no" are important: they eliminate the obviously flawed from the subtly flawed.[3]

To appreciate the power of this principle, recall that all living things ask questions of nature, with their lifestyles, their living worldviews. When nature answers "no," they die.[4] A billion years of such questioning resulted in the enormous diversity of life on earth. The flip side is that most living worldviews, from trilobites to dinosaurs, eventually fail.

This analogy highlights the cardinal weakness of scientific conversations: every theory we know is sometimes violated because observations exist that contradict it or are beyond its grasp. Put bluntly, every law of nature we know is false.

In the brief history of natural sciences, we have already rejected almost all scientific explanations ever created. From this perspective, our explanations of the world appear infinitely more fleeting than the living worldviews of other species, which may persist for millions of years. Many scientists know this but find it too bleak to live by. And so we stick to the rule of "law" but allow that the rule admits exceptions. To do so, however, is to erase any hope of understanding the world completely and with certainty.

A metaphor that dominates the thinking of many about science, the metaphor of progress, can shed further light on these limitations. Does not science creep more and more closely toward a final truth—toward reality? Current scientific theories appear astonishingly successful, perhaps much more so than earlier theories. Not only do they explain phenomena that our immediate experience cannot reach, such as the temperature inside a star or the amount of matter necessary to make one. They have also radically changed the world. They allow us to design machines that fly, produce medicines that cure countless diseases, build computers that grind through mountains of data, and establish global communication networks that carry torrents of information. Scientific explanations work. Powerfully. It is hard to argue with this extraordinary success. But does that mean that we are creeping ever closer toward reality?

In the metaphor of *approaching* reality, the image is one of walking on a path toward a destination.[5] We begin at one end of the path, the point of absolute ignorance, and approach the other end, reality. As we do so, our distance to reality diminishes. But this image, straightforward for every day life, has serious problems. Who or what is absolutely ignorant? An infant before it learns to babble? Infants already have powerfully successful interpretations of the world carved into their minds and bodies. A society ignorant of scientific conversations, perhaps a preliterate society? Hardly. Every human society has the knowledge necessary to protect itself against the elements, to cure diseases, and to feed itself. Are perhaps nonhuman organisms absolutely ignorant? No luck here either. All organisms, from bacteria to chimpanzees, have knowledge: powerful interpretations of their world that work. We know this because they have made a successful living for much longer than we humans. Thus, the point of departure for the journey toward reality is elusive. We will certainly not find it among living organisms.

How far away are we from reality, the other end of the path? If reality is infinitely far away, we cannot possibly approach it, no matter how far and long we travel. So let us suppose that the journey actually has a reachable end point. What does it mean to approach that point? Our distance to it should become smaller. If so, we should be able to gauge that distance. But because we do not (and perhaps cannot) know where the end point of our path is, we have no way of telling how close we are to it. Even if we pursued the more modest goal of comparing two scientific theories to see which is better, we would be at a loss because nobody has successfully counted the number of observations a theory — any theory — can explain.[6]

In sum, we can not say how far we are from the starting point of our journey, how far away the end point is, and how far each step takes us. The everyday image of *approaching* reality fails completely. This will disturb us only if we insist that our scientific journey must end somewhere. It leaves us unfazed if we accept theories as metaphors. They may be the most sophisticated metaphors ever created, but they will occasionally fail — as do all metaphors.[7]

In light of these limitations, the perspective I suggested earlier may no longer seem so far-fetched: the edifice of human knowledge resembles an unending, ever-growing maze with no exit, an endless cabinet of mirrors, in which each reflection mirrors countless others — the interpretations of the world it is built on.

Repeated Questioning

A second pillar of science is the requirement to pose the same questions more than once. If a theory's predictions hold only in one experiment but not in several other identical experiments, the theory is rejected. Theories have to prove themselves under repeated questioning.[8]

But what does it mean to ask *the same* question repeatedly? Every experiment is performed in a (slightly) different world. To paraphrase the

Greek philosopher Heraclitus, you never step into the same river twice. To ask the same question twice is thus strictly impossible.

To address this problem, a scientist divides the world into two parts, the part that influences a phenomenon of interest and the part that can be safely ignored. The part that influences the phenomenon has to be precisely controlled. Take again the question whether *Penicillium* kills bacteria and the experiment that asks this question. The position of Venus relative to Mars, who governs the country at the time of the experiment, the price of gasoline, and whether it is day or night may matter little. But fail to control an important variable, say the species of bacterium you study, then your experiment has contradictory outcomes: the question may sometimes be answered with a "yes" and at other times with a "no."[9]

What determines whether water occurs as a liquid, ice, or vapor? Temperature and pressure. Whether a comet careens into the earth? Its mass, velocity, and trajectory. Whether a food molecule binds a receptor? The molecule's shape and electric charge. Whether an enzyme enables a chemical reaction? Its shape, the ambient temperature, and the amount of salts and ions around it.

As these examples illustrate, laws of nature involve typically only few variables, such as mass, temperature, or electric charge. Part of the reason is that we prefer simple laws and that our minds' ability to grasp relations among many variables is puny. But more important, to ask the same question twice — even when applying "sameness" liberally — is impossible if too many factors matter to the answer.

This is the core weakness that comes with requiring repeatable questions: to identify a few factors that need to be controlled and would allow us to ask the same question twice is not always possible in the real world. The best examples come from human society. We may have no general theories to explain or predict the gyrations of financial markets, the outcome of elections, or the ebb and flow of fashions because we cannot isolate a mere few variables influencing them. Millions of people may matter for these phenomena, their interactions and preferences, the organizations they form, and how these organizations influence one another. With that many factors, it may be impossible to ask the same question twice.

The same holds in the nonhuman world. Whether a swimming bacterium changes directions depends on the pushing and shoving of innumerable particles, atoms, molecules, each of which may cause its rotating flagellum to change direction. Wherever many factors matter to a phenomenon, wherever we cannot ask the same question twice, simple laws of nature elude us.

Systems with this property are often called complex systems and consist of many interacting parts.[10] The interactions are often simple, such as two particles bouncing off each other. But put thousands of these parts together, whether molecules inside a bacterium, ants in a colony, or people in a society, and simple laws often disappear.[11]

Over the past few decades, it has become clear that complex systems are everywhere. They include the water vortex in your emptying bathtub, the movement of a falling leaf, an ant colony in your backyard, the amount of traffic on a road, and the housing prices in your neighborhood. Even physics — most successful in finding elegantly simple laws of nature — is becoming increasingly dominated by complex systems. Prominent scientists such as the astrophysicist Stephen Hawking have gone so far as to proclaim the twenty-first century the century of complexity.

With the rise of complex systems, the unpredictable, individual, and historical regain prominence in our world. Because scientific conversations rely on repeatability, they have no stake in the unique; they leave the individual untouched.[12] This is the shadow cast by science's need for repeated questioning.[13] Together with the absence of certain and final knowledge, it forms the two limitations that follow from the strengths of science. And beyond these limitations is a third, perhaps even more profound one, a limitation about the nature of truth itself.

Ambiguous Truths

Ending a scientific conversation, declaring victory or defeat in explaining an observation, is often a matter of choice. However, a few exceptional conversations have smashed into obstacles as solid as brick walls. These obstacles are paradoxes that occur in the hardest of hard sciences: physics and mathematics. Nothing underscores better that paradoxes are not mere language games. I have already alluded to paradoxes in physics and will now turn to mathematics.

One can prove beyond doubt that some mathematical paradoxes are irresolvable. Such proof is the privilege of mathematics. It cannot be expected for paradoxes in other sciences, such as biology. However, consider that aside from this difference, the paradoxical tensions we encountered in the living are not fundamentally different and are certainly no less fundamental.

You might think that mathematics exists entirely within our minds rather than that it emerges from a conversation with nature, but it is worth keeping in mind that mathematics is the language of choice for most theories, whether physical, chemical, or biological. These laws govern everything from bouncing molecules and the chatter of nerve cells to spreading diseases and spinning planets. Mathematical laws profoundly reflect fundamental properties of our world.

The foundations of mathematics itself are axioms and rules of inference. Axioms are self-evident mathematical truths. For instance, one of the five central axioms of geometry is that you can draw one and only one straight line between any two points.[14] Another axiom, from a branch of mathematics called number theory, is that for any integer number, however large, there is a number that is larger by one, its successor.

Rules of inference are instructions to infer new and true statements from already accepted true statements. Most such rules tend to be simple and intuitive. Take the rule called double negation. It states that the opposite of the opposite of a statement is the statement itself. With this rule, the statement "It is false that I will not go shopping today" reduces to "I will go shopping today."

Beginning from axioms, mathematicians use rules of inference to prove theorems, true — but not obviously true — mathematical statements. Here is an example: there exist positive integers x and y that, if squared and added, are equal to the square of another positive integer, z ($x^2 + y^2 = z^2$). Proving this and other theorems means reducing them to axioms via rules of inference.

Again, it is important to keep in mind how successful this business of mathematicians is: the laws of nature that allow us to build airplanes, computers, skyscrapers, cars, and so on all rely on mathematics. The truths of mathematics do not just exist in our heads. They have material consequences.

Some mathematicians try to find axiom systems, small numbers of axioms from which all theorems of mathematics follow. (Their quest is the same as that of other scientists, to reduce the complexity of the world to a few simple principles.) An axiom system must achieve two goals. It must be able to prove all — known or unknown — true theorems. And it must be free of contradictions. Mathematicians call such an axiom system *consistent*. The opposite, an inconsistent axiom system, would allow you to prove both a theorem and another theorem that contradicts the first theorem. Put bluntly, inconsistent axioms are useless because you might use them to prove anything.

The big problem with inconsistency is that it need not be obvious but may be a subtle flaw, like a microscopic hairline crack in the foundation of a mile-high skyscraper. Initially, as you build floor upon floor — theorem upon theorem — your edifice seems rock solid, as if made for eternity; then, just before you add the last floor, the skyscraper may suddenly cave in and fall to pieces.

The quest for a small and consistent axiom system was an important part of mathematics in the early twentieth century. For some time, it also seemed promising. For example, the mathematicians Bertrand Russell and Alfred North Whitehead invented — or did they discover?[15] — a good candidate for such an axiom system. This axiom system applies to most known mathematical theorems, and it consists of a mere handful of axioms.

It may seem a small step to go from axioms for most *known* theorems — which may already number in the millions — to axioms for *all* theorems. But in 1931 the Austrian mathematician Kurt Gödel dealt a devastating blow to this endeavor. He proved a theorem that denies the existence of such an axiom system, and he did so with a clever trick that allows a theorem to talk about itself. His proof is complicated, but the main idea is easily explained.[16]

Gödel's devastating theorem is akin to the statement "This sentence is false," a statement that harbors a famous paradox called "the liar's paradox." For if this statement is false, then it is true; and if the statement is true, then it is false. Gödel's theorem makes a similar statement about itself. In the arcane language of mathematics, the theorem essentially says, "This theorem cannot be proven." If it can be proven, the theorem is false, and one has used axioms to prove a wrong theorem. Thus, the axioms are inconsistent. Anything built on them will eventually collapse. But conversely, if the theorem cannot be proven, it is true. In that case, the theorem is true but cannot be proven. It is not that we do not know the proof. We know that there can be no consistent proof.

In sum, any axiom system such as the one that builds the mathematics we know must be either inconsistent — and thus useless — or incomplete, meaning that some mathematical truths are unknowable. You cannot have both.

Gödel's insight leads to even more serious problems. For example, the theorem "This group of axioms is consistent" may itself be true but un-provable.[17] This means that the countless theorems on which our laws of nature rest may be riddled with contradictions, although we may never find them. Gödel's devastating insight gave rise to a desperate search for remedies, but none could be found. Not only that, similar paradoxes were found elsewhere in the foundation of mathematics.[18]

Paradoxes like "This sentence is false" show that access to truth is limited — even if we consider only mathematical truth. And again, because mathematical truth is the basis of all science and technology, such para-doxes are not just language games. Their material consequences are best demonstrated with an example from computer science: one can build a machine — a computer — that embodies one of them.

The Paradox in Matter

The famous mathematician Alan Turing envisioned computation as akin to such human arithmetic operations as counting and addition.[19] His notion of computation is still taught to most computer science students. Turing searched for the simplest possible machine capable of performing any computation, and he found it in 1936. Now known as the Turing machine, it is a very simple device, so simple that you could build it at home. It uses only three ingredients: a (long) roll of paper to write on, a small number of symbols, and a few instructions.[20] Yet this machine is so powerful that it could compute everything a modern digital computer can, and more. Turing's ideas greatly influenced the subsequent design of digital computers.

Turing machines capture computations that involve few discrete and distinct symbols, such as the ten digits in our decimal system. Before saying more about them, let me note that not all computations are like that.[21] Think of a colony of honeybees that "calculates" how many workers will be dispatched to a flower or an ant colony that "calculates" which hole in the ground is large enough to house the colony.

If organisms compute differently from digital computers, we may need a more general perspective on computation, such as the following: a computer takes something, its input, and transforms it into an output. The transformation itself is the computation. In this perspective, computation is everywhere. Take a flying bird approaching its nest. Its brain is a computer with goals similar to those of an airplane's flight computer. Its input is position and speed, and its output ensures that the bird lands safely.

The necessary computation, however, involves not only the brain but the entire body. Nerve cells, located everywhere from the tail to the wingtips, help adjust a bird's position when a sudden gust of wind arrives. Similarly, the input that eyes provide to the brain is the result of an extremely complicated computation that begins at the retina, itself a computer. This computation constructs the objects around the bird, including the nest itself. You could thus view the whole bird as an embodied, flying computation. Swimming bacteria make a similar case. Their input is the information embodied in the food molecules available at any one time. The output is a decision: where to swim.

From this perspective, just about any organism's behavior is based on computation. More than that, computation is not even limited to the living or the human-made. You can view two rotating bodies such as earth and moon as a computer. The input is position and speed of the earth and moon at some time. Its output is speed and position at some later time. The dripping of water from a faucet, the swaying of trees in a breeze, and the combustion of fuel are all computations — if a mind creates meaning in their output.[22] In fact, many years before the age of electronic computers, humans had already built sophisticated miniature machines — now called analog computers — that emulate some aspect of the world, such as the motion of planets or the rise and fall of tides.[23]

Turing machines are thus only one of many kinds of computers. They provide a limited perspective on computation. However, this perspective is rich enough to illustrate that paradoxes do not live only in language. The reason is that one can build computers that can prove mathematical theorems. Alan Turing himself showed how to build a machine — an object as tangible as a skyscraper — whose sole purpose is to prove theorems.

Not only that, he proved incontrovertibly that no other computing machine can be more powerful than his.

The input of the theorem-proving Turing machine is a theorem. Its output consists in a decision, the decision whether the theorem is true or false. Given a theorem, a Turing machine churns away as long as it needs to arrive at proof or disproof.

A key question about this machine is whether it can prove or disprove *any* theorem.[24] Alan Turing answered the following variation of this question: Can one be certain whether the machine — given an arbitrary theorem — will ever halt, either returning "true" or "false"? If not, we cannot be sure that the machine can prove or disprove every theorem, because it might simply churn away forever at some theorems. This question is known as the halting problem. It is where paradoxical theorems — in particular Gödel's — come in.

Turing showed that the halting problem is unsolvable. The reason is similar to the reason why the sentence "This theorem is unprovable" leads to a paradox. Turing showed that you cannot build a computer that can prove all true theorems in a finite time. In other words, any one computer might never stop for some true theorems even were it allowed to work on a proof for longer than the universe is old.

In sum, while you could argue that paradoxical theorems exist only in people's heads, a computer, a tangible piece of matter that starts out proving a theorem and never stops is a different affair.

An Endless Road

The following analogy provides another useful perspective on the paradoxical tensions we have encountered here and earlier. Actually, it is more than an analogy; it is an object you can make and explore yourself — another metaphor that does not just live in language.

Take a sheet of paper and cut off a narrow strip, perhaps a centimeter wide and some twenty centimeters long. If you glued the two narrow ends of this strip together, you would turn your paper strip into a paper loop. If you turn one of the ends by 180 degrees before gluing them, however, you have created a much stranger object. It is called a Möbius strip.[25]

This strip has some remarkable features. Run your finger along its surface. Is your finger on the inside or on the outside? You may have started on the inside, but after you have gone around halfway, you will be on the outside. If you have started on the outside, you will eventually end up on the inside. This strip, if you look at it as a whole, has neither an inside nor an outside.

Now pretend that you are small and myopic, that you are standing on the strip, and that you can view only the tiny portion of the strip that you stand on — like most of us whose minds grasp a tiny aspect of the world. It will be very clear to you what side of the strip you are on, inside or out, right or wrong. But if you travel far enough on the strip, you will, without noticing, have changed sides. What is more, if you deform the strip you can shift the region where inside changes into outside.

The same holds for the paradoxes we have encountered here. You can confine them so well that they would be invisible from your vantage point. But they are like a pebble in a boot. During a long walk, they may wander around, may be bothersome at times and invisible at others. But they will not disappear. Eventually they will make themselves painfully known, as much as you try to ignore them.

In physics — the rigorous science of matter — phenomena such as the wave-particle duality and the uncertainty principle have paradoxical consequences.[26] Mathematics — the most rigorous science of the mind — contains paradoxes about truth and falsehood, the two "elementary particles" of abstract thought. Superficially, these paradoxes are not alike, except that both show how closely linked mind and matter are. For mathematical theorems, the link emerges from a computer that can be built to prove theorems. This computer may get entangled in its search for truth and stumble deeper and deeper into an endless labyrinth. Conversely, the paradoxes of physics involve an observer and its interaction with matter. For why is it impossible to determine position and momentum of a particle simultaneously? Because the observer interferes with the particle. And what makes a particle appear as a wave? The question an observer has asked.

The paradoxes of physics and mathematics are easiest to accept because of the rigor of physics and mathematics. But the paradoxical tensions we encountered throughout, tensions between self and other, part and whole, safety and risk, creation and destruction, and so on, are just as substantial, even though we may never have rigorous proof of them. Similar to mathematical paradoxes, these tensions relate to truth and falsehood: they regard "true" ways of looking at the world, "true" answers to simple questions, such as whether organisms act on their own behalf or on that of others or whether the parts or the whole are responsible for bacterial swimming.

These paradoxes are also the most impressive showcases for our mind's tremendous power, a power so great it can examine itself and touch the walls of its prison.

This brings me to a last question: Where are our scientific conversations headed? Without doubt, they have enormous power, which rests both on their rules and on the creativity of mind, the power of creation itself. They have already transformed the world we live in like nothing else before them. And they will transform it further in ways I do not dare to predict.

I do, however, dare to predict something: that any incisive scientific conversation will arrive at the brick wall of fundamental paradoxes. It will arrive at the turning point where wrongs change into rights — like a journey on a Möbius strip — even though the journey may be longer than a human lifetime.[27] The ambitious theories physicists dream about, theories of the origin and fate of the universe, are not exempt. They might be cast in arcane language, intelligible to only a handful of people. But at their heart will be a simple statement about the world, a statement that reflects the fundamental paradox of existence.

In all unending debates of science and philosophy, some of which have been going on for a thousand years, paradoxes play a key role. Consider that these paradoxes will not go away; that, in blinding ourselves to them, we simply continue our ancestors' time-honored strategy; that paradoxical tensions such as those we have encountered are integral to the world and permeate it like pores permeate a sponge.[28] Fundamentally they reflect just one tension, reflect it millionfold, a tension at the root of everything. The tension is that between the opposite sides of a coin, completely opposite yet inseparable, a tension about which Heraclitus long ago remarked: "All things proceed by strife."

CHAPTER 10

The Power and Burden
of Freedom

Man is responsible for what he is. . . .

Man is free, man is freedom. . . .

Man is condemned to be free.

Jean-Paul Sartre

Why is there something and not nothing? What is our place in the world? What is our future? Over two thousand years these questions germinated and grew into Western philosophy. Philosophy, in turn, has given rise to the natural sciences. And even though science has taught us much about the world, it cannot solve the elementary tensions I have touched on in this book. The questions involve the relations between part and whole, self and other, and matter and mind, for example. These tensions turn up everywhere science has traveled, from atoms to organisms and their societies. And they lead to the same paradoxical place, in which a loop of thought apparently allows no escape. Despite their superficial differences, these tensions are all facets of the same principle. We will now briefly revisit some of these tensions and then discuss the burden they levy on humans: the burden of freedom.

Revisiting Wholes and Parts, Matter and Mind, Self and Other

Part or whole, which is responsible for shape-changing receptors, swimming bacteria, or the making of an organism? Scientists are in the business of taking wholes apart — in the laboratory or in their minds — and rebuilding them after studying the parts. This practice is essential for science and it is one reason for its success. Some scientists, however, have made a leap from this essential practice to a dogmatic principle: to understand the parts is to understand the whole — there is really nothing but parts. Not surprisingly, their vigorous defense of this perspective triggers a backlash, mostly among nonscientists. In this backlash, many reject scientific practices altogether and simply claim that the whole is more important than the parts. They reject scientist's contributions to society, contributions such as "conventional" medicine, genetically modified foods, and nuclear power.

Each camp claims its perspective as the truth, as how the world *is*. But these perspectives are complementary. Either one is incomplete, for parts and wholes are as two sides of one coin, completely separate — opposites — but at the same time inseparable, one molding the other. Nevertheless, it is nearly impossible to hold both perspectives at the same time, just as one cannot stand in two places simultaneously. For any scientific conversation, indeed for any question one asks, it is necessary to choose one of these perspectives, to focus either on the whole or on some of its parts. This paradox of whole and part is as fundamental as the famed paradoxes of physics.

In the face of such a paradox, power derives from choosing one side and sticking with it through an inquiry, a conversation with nature. There is just one caveat with such choice: to forget that any choice is just that — a choice, and no ultimate truth — will eventually prove fatal. It will eventually stop any inquiry dead. Any choice is meaningful only in the context of a conversation. Put differently, the choice is always between two perspectives and how they answer a question or propel an inquiry further. Hanging on to a choice as a final truth will ultimately be a roadblock for any inquiry.

To accept a paradoxical tension like that of part and whole as fundamental is to facilitate choices of perspectives that advance a conversation. Awareness of a paradox breeds the power to choose. This choice's power is the power to create, to formulate powerful worldviews. Choices of perspective truly have power? Yes. Just recall where the choices of scientists have led in as little as a century: to technology—for better or for worse—that would have appeared sheer magic to our ancestors. They have changed the world.

This one tension, between part and whole, already permeates the world. But many other facets of the same principle are equally pervasive. One is the tension between matter on one hand and mind or meaning on the other. This tension is old. Scholars have argued about it for hundreds of years. Many current thinkers tend to emphasize one of its sides: matter is everything. An example is the view that mind is *nothing but* a by-product of electrical discharges in the brain. From this viewpoint, our conscious experience is only a by-product of atoms bouncing into one another, an illusion of humans looking at themselves, not part of a world "out there." But careful inspection unveils meaning everywhere in and around the living, from humans to bacteria, from whole organisms to molecules. Communication and flow of meaning are ubiquitous, albeit often in strange and unfamiliar forms. Even molecules are exchanging signals by changing their shapes. Meaning is literally everywhere. To be sure, it is always associated with matter. But because of the ubiquity of meaning, the converse perspective, that meaning is primary to matter, is equally defensible.

Present-day computers reflect perhaps best the prevailing perspectives on matter and meaning as well as their shortcomings. Most people conceive of computers as consisting of two utterly different parts. One is the hardware, such as transistors and wires, and the other is the software, the instructions that endow the computer with its "mind." The world, of course, limits the extent to which hardware and software can be separate. Meaningful instructions are stored inside hardware—meaning thus depends on states of matter. And conversely, these states of matter depend on meaning, for the hardware undergoes subtle changes—changing flows of electrons or opening and closing switches—because of instructions in the software. Nevertheless, hardware and software are usually viewed as utterly separate.

Compare this with the many ways in which nature computes. Nature builds vastly successful computers, implemented as nervous systems and brains, but also as objects as different as flying birds and bacterial cells. In a brain, billions of individual nerve cells are connected through living wires.

Electric signals that crisscross this neural network perform sophisticated computations. Arguably, nerve cells and wires are the hardware of these living computers. But where is the software? Nowhere separate from hardware. Hardware and software are one and the same. The network itself determines the kind of meaning it can recognize, digest, and generate. And conversely, the computations a living computer can perform, its instructions, determine how nerve cells must be wired to carry them out.[1]

A third paradoxical tension reveals itself in the relation between self and other. Many think of different organisms as completely separate, each struggling fiercely to further their own interests. But organisms are often related through a common fate, sometimes mediated through shared ancestry and common genes, sometimes through other dependencies — shared food, shared living space, and so on.

A paradoxical tension arises because organisms are linked in their fate from one perspective yet are completely separate from a different perspective. We cannot do justice to both perspectives at once. This tension leads to bitter struggles: controversies about whether organisms ever act on behalf of other organisms or whether, ultimately, every organism is in it for itself. But the argument evaporates if one appreciates that either perspective is incomplete. Which to choose depends on the questions one asks. And making this choice, in awareness of its inadequacy, avoids entanglement in a paradox.

These and other tensions I have emphasized here — self and other, matter and meaning, part and whole, safety and risk, creation and destruction — come from the world of the living. Yet they are every bit as profound as the similar tensions in mathematics and physics we discussed in the previous chapter, tensions that create fundamental limits to our knowledge. My modest selection of examples does not do these tensions justice, for we are truly immersed in them.

One and the same object harbors many of these tensions simultaneously, even apparently simple objects like receptors for food molecules: they and their amino acid parts already harbor the tension between matter and meaning, part and whole, and creation and destruction. And each of the two opposing perspectives such tensions generate is incomplete without the other, even though we cannot see the world both ways at the same time. To escape getting entangled in paradox, we have to choose a side. Such choice is at its most powerful if we are aware of choosing, if we do not confound this choice with an ultimate truth. Because any such choice is justified only through the conversation it is a part of and how it propels that conversation onward through the questions it answers.

The Paradox in Human Life

Given the ubiquity of paradoxical tensions elsewhere, it is not surprising that they also arise in the choices humans make every moment to steer their lives. Some choices are of course easy. But all humans occasionally face an impossible problem, with no best solution, and only imperfect options. As when facing a paradox, they then have to choose despite the near impossibility of doing so. (It requires strength to act in that situation — but perhaps less if one believes that there *is* right and wrong given to us from a place beyond us.)

Many difficult choices — whether of individuals, groups, or entire societies — bear the mark of the paradox, namely that any choice violates truths held as fundamental. In a complex society governed by multiple institutions — a parliament, a head of state, a judiciary — it is easy to overlook this paradox. Nonetheless, it is omnipresent. I will let one example stand for many others.

A fundamental problem any government faces is known as the "paradox of democracy," although it applies to any form of government.[2] The paradox of democracy consists of the observation that commitment to a democratic government harbors the seeds of its destruction, because a democratic government can vote itself out of existence. The People could decide that a tyrant is better suited to rule than the People. Democratically elected dictators such as Hitler illustrate this paradox.

Many other fundamental societal values reflect the same paradox. A wholly tolerant society harbors the seed of its destruction, because the tolerant will tolerate the intolerant — fanatics, zealots, and hate mongers of all stripes — who ultimately will do away with them. How to maintain democracy and tolerance yet shelter them from destruction? The question has no clean answer. It reflects a paradoxical problem, which societies can escape through a choice that violates the core principle of democracy. The choice of most democratic governments is that the tolerant must suppress the intolerant, by force if necessary.[3]

The paradox of democracy also reveals itself in more subtle forms.

Modern democracies soften its impact by establishing balances of power among separate institutions. Despite such safeguards, the paradox breaks to the surface occasionally, and nowhere more dramatically than in constitutional crises. In the United States, for instance, President Richard Nixon challenged the balance of power when he threatened to resign only on a "decisive" vote by the Supreme Court — although the decision to call a vote decisive would have rested with him alone.[4] This balance has also been challenged in hung elections, such as the U.S. presidential election of 2000, whose outcome had to be decided in court. And it is constantly challenged through the evaluation of the constitutionality of laws. According to the U.S. Constitution, new laws ultimately come from the populace. Yet the decision whether new laws are in agreement with the most basic of laws, the Constitution, is in the hands of the Supreme Court.[5]

The paradox of democracy is mirrored by a fundamental paradox faced by every human, the paradox of freedom. That is, an individual's freedom to choose includes the freedom to abandon this freedom. The example of *Moby-Dick*'s Captain Ahab illustrates this choice. When his leg is severed by a whale, Ahab asks the ship's blacksmith to cauterize the stump with a red-hot iron and relinquishes his choice to stop the blacksmith from doing so. During the grueling procedure, Ahab screams in anguish and pleads to be spared, but to no avail. He has abandoned his freedom to choose.[6]

Although little appreciated, the paradox of freedom is part of everybody's life. It is embodied in a cornerstone of any society, the enforceable contract, which is a promise of one human to another, a company, or the government. It is a part of every irrevocable commitment a human makes to another human. Choices that cannot be reversed easily, such as enlistment in the military or indebture, illustrate it best. The most drastic irrevocable choice we can make is the choice to end one's life. It is no surprise that the philosopher Albert Camus called suicide the only serious philosophical problem.[7]

Disorientation and Seekers of Paradox

If you become skilled at following questions until their paradoxical end point, you may begin to see that paradoxical tensions are truly everywhere. You may come to watch yourself make choices where previously only truths reigned. And the moment you become aware of choosing, of realizing a choice for what it is, the solidity of a truth may evanesce into thin air. Your mind may begin to feel like a stray traveler in a featureless desert, with only shifting sands beneath you and no solid ground to stand on, except when you choose to rest on a truth, exhaustedly perhaps. You may find yourself walking on a dangerously exposed path, where any gust could prove fatal, where any wrong decision, any misstep, could make you lose yourself.

Why is it so maddeningly difficult to focus the mind's eye on paradox? The ancestors of humans — all the way back to bacteria — may hold the answer: the history of successful life is one of commitment to *one* worldview, whether crudely embodied in a living cell or more subtly in human thought.

Imagine a monkey swiftly swinging from tree to tree. The monkey sees a snake on the ground, climbs down to the snake, and ponders the significance of the encounter. Snap — the snake strikes. The monkey is dead. The executive of its own life, the monkey must rely on quick decisions and actions to survive. For most organisms, to abandon their embodied worldview is to invite death. Blindness to potentially paralyzing paradox thus comes very naturally to the living.

To portray *all* living beings as avoiding paradoxes would be misleading. Some humans have been aware of its importance since the dawn of civilization. The earliest records in the West come from the Greek philosophers. Heraclitus's statement that "all things proceed by strife," or Socrates' famous sentence "I know that I don't know" were among them. Did the ancient philosophers already see that the paradox is like a crack in the world-egg through which creation continually struggles to hatch? We may never be sure of that, but we know that paradoxes have played a central role in some Eastern traditions. They are used to help an apprentice glimpse the place beyond paradox, beyond meaning and matter, self and other, part and whole. To this end, he or she would be given an unsolvable paradoxical riddle, such as "What is the sound of one hand clapping?" The apprentice would then commit years of hard work to solving the riddle. The solution arrives when the disciple — often exhausted — loosens his or her grasp on the effort. And then the apprentice realizes that the solution, like the paradox itself, is already everywhere. About this process, the masters would say, "Enlightenment comes when the way of thinking is blocked."

These approaches to paradox may be profound, but they play a minor role in the history of thought. This is not surprising, especially given the rise of natural science in the past centuries. Science's fantastic success inspired devotion to an objective reality and to ultimate truths. Paradox is the mortal enemy of these idols.

For some time it seemed as if science might not only solve all the world's problems but also provide a complete and seamless explanation of everything. Yet it also deeply alienated humans from the world they had created. By the end of the twentieth century, it was clear that science would fall short of the goals humans had set it. For one thing, science would not end injustice, wars, and general misery. People and their choices need to do that. But even more important, fundamental limits of scientific knowledge are now evident, fundamental paradoxes in science's foundation, theorems true but unprovable, computers that should but may not halt, particles that appear like waves, and so on. Although science has created the most powerful conversation short of life itself, it also has deep-seated limitations.

Why Live with Paradox?

If it is difficult to see the paradoxical tensions all around us, why make the effort? Isn't it much better to crawl back into the prison of final and certain knowledge, to leave behind an open world? What would we gain by appreciating paradox and the limitations it implies?

First, there are the benefits of humility and serenity that come with abandoning final truths. To appreciate our limited power might alleviate some human hubris. It might also lead us to take ourselves less seriously and to stop criticizing or persecuting others for their different perspectives. Furthermore, accepting our role might silence some of the human restiveness that comes with feeling alien to this world. We might also appreciate that our place in the inner dialogue of creation is as good as any other place, our role in the world as good as any other role. There is nowhere else to go. This is it.

Second, paradox creates choice in building the world's foundations. When we realize that ultimate truth is elusive, we begin to appreciate that laws of nature are just extremely sophisticated metaphors. Like any metaphor, they resemble what they stand for, yet they differ in key respects.

If all laws and their building blocks are metaphorical, why not begin to build perspectives around different metaphors? Why, for instance, not build a worldview around the notion of meaning, including the meaning that humans perceive and create, to the meaning that molecules carry? Such a "logocentric" worldview would turn the world from a hollow shell into a place teeming with meaning.

The value of such a choice depends, of course, on the questions it can help answer. This choice might, for instance, prove useful in a theory of communication and information, both of which do not come naturally to a worldview based on matter.[8] Concepts like mind and intelligence might fall in place effortlessly, as they do not in a material worldview.[9] (Critics would of course say that a logocentric worldview cannot reflect the *true* nature of mind, but doing so merely exposes their unstated choice.)

Any such worldview is just a perspective with its own blind spots. For instance, explanations of material phenomena — the swinging of a pendulum, the half-life of a radioactive atom, the boiling point of water, or the products of a chemical reaction — might not come naturally to it. Fortunately, physics and chemistry have amply taken care of these phenomena.

Only time will tell whether a logocentric perspective will lead further than alternative possibilities. The central point, however, is that humans play pivotal roles in science's conversation: for humans are the ones choosing metaphors, an activity more akin to that of artists, who seek truth in elegance and beauty, than to that of soulless servants of a hollow god. In this respect, once again, the paradox humanizes science.

Last and most important, *awareness of the paradox returns to humans a great power and responsibility: the power and responsibility of actively participating in the conversation that creates their world.* All of the ever-refined perspectives that emerge from science's conversations depend on a human mind and the meaning this mind creates. Any point in this conversation allows, yes, requires, choices about the next question. Each choice depends on a perspective taken, and each will lead the conversation to vastly different places. Humans are central to this conversation. They are not the insignificant cogwheels of a monstrous machinery they have invented themselves to be.

This does not mean that human power is limitless. The world does not just exist in people's heads — or in your own head, as the solipsists argue.[10] Humans are merely one partner in a vast conversation. The choice to jump out a window based on the perspective that gravity is an illusion would lead to a rude — or no — awakening. A choice is not necessarily as good as its opposite, for every choice is made in the context of a conversation. Nature is our partner in scientific conversations. And nature's answer to a question plays a role in our choices, as does the world in which we live. These choices in turn influence how the conversation will proceed; they influence the next questions. Our human power is limited by our conversation partners.

Among all the consequences of acknowledging the pervasive paradox, human power is the most important. Humans become truly human when absolute certainty and absolute truth dissipate. And with the importance of human choice comes the ultimate freedom, the freedom to create one's world out of conversations with it. Living in the paradox is the ultimate luxury. It is a place where humans are invited to make a difference — every moment until their dying breath; a place where human choices can initiate conversations that can change the world; a place where no committed human effort could be dismissed as quixotic, futile, or crazy. In a thousand years humans might have conversations with the world around them that we cannot even imagine. Human choices may open doors to a near-magical realm of possibility.

The Solitude of True Choice

And yet, all this freedom and power have a consequence as unsettling as the confusion paradoxes cause. For if we have the power of true choice, then nobody and nothing stands behind us. We cannot fall back on anybody, cannot rely on any ultimate authority. We are the ultimate authority. We are alone. And our responsibility and solitude in choosing are a heavy burden. The image of the judge condemning a prisoner to death comes to mind. How can any one human, or any group of humans, take the responsibility of destroying what they have not created? Yet humans are condemned to make such choices, have made them since time immemorial, and will make them forever. And these choices are asked not only of humans in positions of power — judges, lawyers, doctors, politicians, generals. They are part of every relationship — whether among employer and employee, parent and child, salesclerk and buyer, or husband and wife. Any human capable of influencing the fate of others and him- or herself —

thus, every human — has to make them. They are part and parcel of being alive. And every choice is ultimately made in utter solitude, whether its consequences are insignificant or immense. In the face of such solitude, the reign of the old idols — whether those of holy writ or that of ultimate truth — seems a wonderful safe refuge. Nobody could be blamed for wanting to return to it.

One might think that we can turn to ethics for help, that ethics could provide recipes to guide our choices. But if the past is any guide, then ethics fails to absolve us of our terrible responsibility, and not for lack of trying. Ethics' most hopeful ideas have failed to provide universal guidelines for choosing. Take utilitarianism. According to this perspective, a choice is good if it measurably benefits a group, whether family, nation, or humankind. But in practice, utilitarianism turns out to be a hollow proposition. For only some economic benefits — money, property — can be measured and counted. Most others are beyond measurement. Additionally, in countless instances the good of somebody is the detriment of another person. A calculus of happiness has proven elusive.

Another once hopeful candidate is Immanuel Kant's categorical imperative. It dictates that choices are good if they could serve as the basis of universal laws, laws that govern societies. But most such laws are by no means natural laws, given to humans from a place beyond themselves. They are created by humans themselves. And their creation often require agonizing struggles and thus the very choices that they are supposed to guide.[11]

The Choice to Choose

Freedom harbors terrible risks, for each of us, those around us, and humankind. Humans have become capable of enormously powerful and destructive choices. They can choose to kill not only themselves but all of human life and that of the planet. Whether they take responsibility for their world is their choice alone, the choice of becoming a cause in matters that concern them. Freedom, by definition, knows no safeguard from complete destruction. Anything that can fly can also crash and burn.

Is it mere chance that we have thus far escaped disaster, or is there rhyme and reason behind it? Is the reason that which few have glimpsed, the glue of the world, that which is beyond the paradox, but from which the paradox emerges, which explodes silently, outwardly into creation, through the centrifugal force of its tensions, a creation in which the paradox is manifested millionfold, in light and darkness, part and whole, self and other, a cosmic cabinet of mirrors? Has the uncreated creation held the world together for eons, prevented it from being swallowed — whole — into oblivion, and does it continue to do so in the age of humans? We do not know but we can hope. This is all the hope one could ask for.

Faced with the choices paradoxes entail, nobody could be blamed for not choosing them. Nobody could be blamed for not choosing freedom. For freedom means having to accept ultimate imperfection. Freedom means to choose in utter solitude. Freedom means eternal uncertainty. It means that the world is ultimately mysterious and enigmatic.

Living in the face of paradoxes means power, but it also means a voyage through a heaving bottomless ocean. This is not easy to accept. Our inquiries are driven by the desire for final answers, but the paradox stands forever in the way of final answers, making any human inquiry as unending as the creation it is a part of. In contrast to Sisyphus, however, who pushes a giant boulder up a mountain only to see it tumble down in an endless cycle of tedious toil, human conversations never return to the same place. They continually create new realms of inquiry and wonder.

The role of humans in making the world through their conversations is adventurous and thrilling. This role makes any human endeavor, including science, a most elaborate, colorful, and fantastic drama. It is a drama with an open ending, for the paradox makes the world a possibility, not a certainty. In this drama you are not a replaceable stagehand. You are the star of the play. And you are invited to take a role, to become a minute part of the grand conversation. It is the only thing to do, if you choose to do anything at all. To see this possibility, of course, requires a choice. Nobody can take this choice away from you, but neither can anybody make this choice for you. It is the choice behind all choices. It is the choice to choose.

The Right and Duty to Live

There would be nothing
Not one buzzing insect
Not one shivering leaf
Not one licking or howling animal
Nothing hot nothing flowering
Nothing frost-covered nothing brilliant nothing sweet-smelling
Not one shadow licked by the bloom of summer
Not one tree bearing furs of snow
Not one cheek colored by a joyous kiss
Not one wing, timid or bold in the wind
Not one curve of exquisite flesh not one singing arm
Nothing free to win or to ruin
Or to disperse or to unite
For better or for worse
Not one night armed with love or rest
Not one steady voice not one trembling lip
Not one unveiled breast not one open hand
No misery and no satiety
Nothing opaque nothing visible
Nothing heavy nothing light
Nothing mortal nothing eternal

There would have to be a human
No matter which human
Me or another
Otherwise there would be nothing
                    — Paul Éluard[12]

NOTES

The following notes are of two types. Some present facets of arguments too convoluted or technical for the text; they expand on simplifications in the text for the benefit of the reader familiar with or interested in a topic. Others suggest avenues of further reading, either to support a point or to let the reader explore a topic further. This list is eclectic — I have not attempted to be comprehensive but rather have chosen works that represent a viewpoint lucidly or provide access to a large body of literature.

Introduction

1. Jacques Monod, *Chance and Necessity: An Essay on the Natural Philosophy of Modern Biology* (New York: Knopf, 1971).

2. From T. S. Eliot's famed poem *The Waste Land,* see Michael North, ed., *The Waste Land* (New York: Norton, 2001). The quotation is a reference to Philippians 4:7.

3. After Ursula K. Le Guin, *Lao Tzu: Tao Te Ching: A Book about the Way and the Power of the Way* (Boston: Shambhala, 1997).

Chapter 1. The Inner Dialogue of Creation

1. Of course, I have no way of knowing what you think, except my experience. So I could constantly become entangled in my own thoughts about what you think. A paradox is very close here. Can you see it? The only way of escaping it is to make a choice.

2. This is the first of many examples that demonstrate how close matter and meaning are in the world. I call these interactions among people conversations — an exchange of meaning through language. But all these conversations also involve

matter, whether through the compression of air molecules as sound waves propagated in speaking, through the motion of electrons in telephone wires, or through the shape change of retinal photoreceptors as light strikes them. Emphasizing the conversational aspect of all this might seem to downplay the role of matter, but it is a counterpoint to the general overemphasis of matter in our culture. What it all comes down to, as further examples will show, is that matter and meaning are inextricably linked.

3. Thomas A. Sebeok, ed., *Animal Communication: Techniques of Study and Results of Research* (Bloomington: Indiana University Press, 1968).

4. For an accessible entry point into such classification and into the history of signs, see Umberto Eco, *Segno* (Milan: Instituto Editoriale Internazionale, Milano, 1973).

5. One could argue endlessly whether there are signs, for example in the English language, that are more elementary than others. If you have a taste for such arguments, Eco, *Segno,* is again a good starting point.

6. Modified from one of the definitions provided in Eco, *Segno,* which is taken from an Italian dictionary of philosophy.

7. There are many subtle distinctions between a sign and its meaning. Unfortunately, there is no widespread agreement about the most appropriate distinctions and their nomenclature. Three common distinctions are made in Eco, *Segno.* The sign itself is called the *signifier.* The word combination "this note" is an example for such a signifier. What this word combination stands for, this note, is the *signified.* It is possible that the signifier signifies something that does not exist in the world. For instance, the piece of text that you are reading right now might not be a note but a part of the main body of the book. (The book itself might not contain any notes, much less this one.) The class of objects in the world that the signifier refers to is the *referent* of the signifier. These distinctions are to some extent ambiguous, which is part of the reason for much disagreement among philosophers.

8. Hilary Putnam, *Mind, Language and Reality* (Cambridge: Cambridge University Press, 1975), provides an inroad into this field, notably through the following two chapters, "Is Semantics Possible?" and "The Meaning of 'Meaning.'" Putnam presents an analysis of word meanings that is useful in its pragmatism. According to Putnam, the meaning of a word like "water" has four complementary aspects: (1) its syntactic categorization (it is a mass noun); (2) its affiliation with a semantic category (it refers to a liquid); (3) stereotypical features attributed to what it stands for (colorless, tasteless); and (4) its extension, loosely speaking what in the world it refers to, a kind of molecule consisting of two atoms of hydrogen and one atom of oxygen. Agreement among people (collective choice) enters critically at least into the third, and perhaps also into the first two aspects of this definition.

9. What do scientific theories stand for?

10. The distinction is due to Peirce. See, e.g., *The Collected Papers of Charles Sanders Peirce,* 8 vols. (Cambridge, MA: Harvard University Press, 1931–1958).

Iconic signs play a key role in the imitative magic of many native societies, where objects take on magical powers through their similarity to other objects, as discussed extensively in James George Frazer, *The Illustrated Golden Bough,* ed. Mary Douglas (Garden City, NY: Doubleday, 1978). This perceived power is also wildly popular in contemporary culture, as exemplified by homeopathic medicine, astrology, and feng shui. For instance, in feng shui, houseplants with round leaves are thought to bestow prosperity to their owner by virtue of the similarity of their leaf shape to the shape of coins: see Karen Kingston, *Clear Your Clutter with Feng Shui* (New York: Broadway Books, 1999). Peirce also points to a third kind of sign, the index, a sign with a physical connection between signifier and referent. Examples include a flag whose movement indicates wind direction and a finger pointing at an object.

11. On a historical note further elaborated in Eco, *Segno,* a perspective that dominated philosophy for centuries held that all signs, including those of human language, have a deep inner nexus to their meaning. From this perspective derive many efforts to scour words for hidden meaning, for example by assigning numbers to individual letters, adding them up, and then relating them to biblical dates.

12. There is an enormous literature on nonverbal human communication. For example, see Judee K. Burgoon, David B. Buller, and W. Gill Woodall, *Nonverbal Communications: The Unspoken Dialogue,* 2nd ed. (New York: McGraw-Hill, 1996); and Michael Argyle, *Bodily Communication,* 2nd ed. (Madison, WI: International Universities Press, 1988).

13. J. Melamed and N. Bozionelos, "Managerial Promotion and Height," *Psychological Reports* 71 (1992): 587–593.

14. A comprehensive albeit older compendium of animal communication is Sebeok, ed., *Animal Communications.* More recent accounts include Jack W. Bradbury and Sandra L. Vehrencamp, *Principles of Animal Communication* (Sunderland, MA: Sinauer, 1998). Authors differ in their view on what distinguishes human and nonhuman communication; e.g., compare Stephen R. Anderson, *Doctor Dolittle's Delusion: Animals and the Uniqueness of Human Language* (New Haven and London: Yale University Press, 2004); and Terrence W. Deacon, *The Symbolic Species: The Co-Evolution of Language and the Brain* (New York: Norton, 1998).

15. One may ask whether the required choices are conscious. I would argue that as long as we do not have general agreement on what consciousness is, we are ill served by insisting on this distinction.

16. There is a plethora of terminology that I have not mentioned. For a brief introduction, see César R. Nufio and Daniel R. Papaj, "Host Marking Behavior in Phytophagous Insects and Parasitoids," *Entomologia Experimentalis et Applicata* 99 (2001): 273–293. It provides examples of some of the processes and signals that I discuss in nontechnical terms, including marking pheromones, kairomones, redundancy, superparasitism, hyperparasitism, and more. Plants can also fend off attacks by producing chemicals that serve as defensive weapons. These chemical signals warn nearby plants, giving them a head start at producing their own

defense chemicals to withstand an insect attack. See R. Karban et al., "Communication between Plants: Induced Resistance in Wild Tobacco Plants following Clipping of Neighboring Sagebrush," *Oecologia* 25 (2000): 66–71.

17. I have deliberately avoided discussing the purpose of some of the signals involved. Does one damaged plant signal the damage to other plants in order to help them avoid the same fate? We do not know, and it is unlikely. The communicative role of plant damage may be entirely unintended. It is thus important to keep in mind that intention is not necessarily a part of communication.

18. Nufio and Papaj, "Host Marking Behavior in Phytophagous Insects and Parasitoids."

19. An overview of a variety of chemical conversations for one insect order is given by M. Ayasse, R. J. Paxton, and J. Tengo, "Mating Behavior and Chemical Communication in the Order Hymenoptera," *Annual Review of Entomology* 46 (2001): 31–78.

20. Annkristin H. Axén and Naomi E. Pierce, "Aggregation as a Cost-Reducing Strategy of Lycaenid Larvae," *Behavioral Ecology* 9 (1997): 109–115; Anurag A. Agrawal, "Phenotypic Plasticity in the Interactions and Evolution of Species," *Science* 294 (2001): 321–326.

21. Ana H. Ladio and Marcelo A. Aizen, "Early Reproductive Failure Increases Nectar Production and Pollination Success of Late Flowers in South Andean *Alstromeria Aurea*," *Oecologia* 120 (1999): 235–241; Agrawal, "Phenotypic Plasticity in the Interactions and Evolution of Species."

22. An overview of the material on planktonic organisms I talk about here can be found in P. Larsson and S. Dodson, "Chemical Communication in Planktonic Animals," *Archiv für Hydrobiologie* 129, no. 2 (1993): 129–155.

23. Many aspects of this process may also be driven by seasonal variation in predator density.

24. From the genus *Daphnia spp.* See Larsson and Dodson, "Chemical Communication in Planktonic Animals."

25. For a review on such communication, see Stephan Schauder and Bonnie L. Bassler, "The Languages of Bacteria," *Genes and Development* 15 (2001): 1468–1480.

26. You can choose to draw a line between conversations and nonconversations at any of the differences between the different kinds of conversations we are talking about, e.g., at the number of participants, the kinds of signals, or the directedness of a signal toward another organism. And for a particular purpose, such distinctions can be very useful. For instance, you may be interested only in conversations where organisms communicate with each other, and where they do so purposely. In this case, the latter distinction of conversations as involving directed signals may be extremely useful, e.g., M. H. MacRoberts and B. R. MacRoberts, "Toward a Minimal Definition of Animal Communication," *Psychological Record* 30 (1980): 387–396, even if only to define the field of study. However, it is not the *one* true definition. It is a distinction useful for a purpose, an expedient choice of perspective. And for our purpose, the broadest possible perspective is the most expedient.

27. Schauder and Bassler, "Languages of Bacteria."

28. This perspective on matter and meaning goes far beyond the mind-body problem as traditionally viewed, but there is obviously some overlap. For a concise overview of prominent views on the mind-body problem, see Jerome A. Shaffer, "The Subject of Consciousness," in Shaffer, *Philosophy of Mind* (Englewood Cliffs, NJ: Prentice-Hall, 1968). The perspective is only superficially similar to identity theory, which posits that mental and body states describe the same thing. The difference is that the nature of the "thing" is different here. Identity theory is a materialistic theory and presupposes the primacy of matter. However, neither matter nor meaning is primary here. The perspective thus perhaps comes closest to double-aspect theories of mind and body that may well have originated with Spinoza. One difference is that I view similarly paradoxical tensions as permeating the world. Also, we are of course in a completely different position from philosophers hundreds of years ago, in the sense that we know of many examples from nonhuman biology that can bolster this position. And having been taught the centrality of paradoxes by hard disciplines like mathematics and physics early in the twentieth century, we may be in a better position to embrace them as a core feature of the world.

29. John W. McAvoy et al., "Lens Development," *Eye* 13 (1999): 425–437; D. Jean, K. Ewan, and P. Gruss, "Molecular Regulators Involved in Vertebrate Eye Development," *Mechanisms of Development* 76 (1998): 3–18.

30. There are many examples of tissues and organs that degenerate when not constantly in communication with their surroundings. One of the best-studied examples is that of muscle tissue, which needs relentless signaling from nerve cells, not only to form, but also not to degenerate once formed.

31. A biologist might object that there are organisms like the nematode *Caenorhabditis elegans* whose cell fates appear to be determined very early in development. However, detailed experimental studies perturbing this nematode's development using cell ablations and genetic perturbations demonstrate ample cell-to-cell communication in developing embryos. And even if this was not the case, then the interior of early embryonic cells — perhaps even the oocyte — must already provide the structure necessary for later development. Structuring of such early cells also requires communication, although in a wider sense that I will get to below.

32. Agrawal, "Phenotypic Plasticity in the Interactions and Evolution of Species."

33. This is perhaps a good place to declare a position on the philosophical position called external realism. According to this perspective, there is a reality independent of our representations. This perspective is often phrased in such statements as, "If all humans or perhaps even all living beings were extinguished, there would still be things like snow-capped mountains — that is, independent of representation."

I subscribe to the notion that we have not just dreamed up the world with our representations. We know, for example, that the world resists many of our representations, as when an aspect of a theory in the natural sciences is rejected by an experiment or when I bang my toe painfully against a wall. But to conceive of snow-capped mountains without a representation of "snow," "cap," or "mountain" is impossible. It is like conceiving of "meaning" without meaning. Any inanimate object that we can talk about is tied to our representation of it. Whatever is out there, there is no way to conceive of it without a representation. It is Kant's famous *Ding an sich*. Whatever is behind the world as revealed through our senses, it must remain forever in the dark. In other words, representations are essential to conceive of the world, and I argue that they play a big role in shaping the world itself. What I am after here, however, is to describe the world not so much in terms of the mind but in terms of a dialectic relationship between mind and matter, one shaping the other.

34. Strictly speaking, cells do this indirectly. Water will tend to flow into the direction that tends to achieve osmotic balance, equal ion concentrations inside and outside the cell. Cells can regulate the direction of water flow by pumping ions into the cell or out of the cell. See, e.g., Lubert Stryer, *Biochemistry,* 4th ed. (New York: W. H. Freeman, 1995).

35. Again, this process may be indirect, for example, it may be facilitated by ion transport.

36. The language of chemistry describes interactions of receptor and signal (or of any two molecules) in statistical terms—that is, through numbers such as forward (association) and backward (dissociation) rates that describe how many interactions are expected to occur on average per unit time.

37. Cells make a response conditional by chemical modification of receptors or by blocking a network of events leading from message to response at one or more essential steps.

38. To ensure that, say, the development of an embryo proceeds in an orderly fashion, many molecular messages are usually sent from one tissue to another. If enough cells respond in a certain way then development will proceed properly, regardless of whether an individual cell responds.

39. Albert László Barabási et al., "Parasitic Computing," *Nature* 412 (2001): 894–897.

40. Thus arise ethical and moral concerns that have crystallized into social scientists' agreements with the world. These agreements discourage inquiries that manipulate people's lives, and they are written into a scientist's code of conduct and into a society's laws.

41. Like any other conversation, this conversation does not start from scratch. I know what a plant is, what a window is, and what a windowsill is. I know where the plant is in relation to the window, what it means for the leaves to be parallel to the window, and so on.

42. Do philosophers have a place in this perspective on science? What is their conversation with the world?

43. Other reasons why the elementary building blocks of physics have at best metaphorical character are emphasized by Roger S. Jones, *Physics as Metaphor* (Minneapolis: University of Minnesota Press, 1982).

44. Another important and relevant insight from quantum physics is the importance of the observer in the state of the observed. The observed does not exist independently from the observer; it is profoundly changed by the act of observation. One may be tempted to discard the notion of the observed and say that there is just this conversation between a human (a physicist) and the world. (What does the world say to the physicist?)

45. Of which many current efforts to let meaning emerge from a mechanistic worldview bear witness.

46. David Hume, *Enquiries Concerning Human Understanding, and Concerning the Principles of Morals* (Oxford: Clarendon Press, 1975).

47. There is of course a need to distinguish different kinds of association, some of which one might call spurious and others that would be better evidence for a causal nexus in a causalist worldview. But strictly speaking, even the latter are still just that, associations.

48. A relevant term used by Eco, albeit emphasizing signs in human language, is "unbounded semiosis" (Eco, *Segno*, section 5.4). In general, pansemiotic worldviews are nothing new in the history of thought. However, they have usually centered around a religious interpretation of signs — namely, that everything in the world points to a creator.

49. To be sure, the language of the world encompasses much more than human language. But because all things meaningful have commonalities, we may learn about this mother of all languages by examining more familiar languages. One such language is the language of literature. What stands for another in this language is metaphor, a sign endowed with meaning. And in a worldview centered around meaning, everything is a metaphor. Everything is related to something else but at the same time also irrevocably separate and different. Just as in a metaphor.

Chapter 2. The Other Side of Self

1. André Frossard, *Forget Not Love: The Passion of Maximilian Kolbe,* trans. Cendrine Fontan (Fort Collins, CO: Ignatius, 1991).

2. Even if the well-being of their children only gives the parents happiness, what parents do for their children could be construed as hedonism and thus, ultimately, nonaltruistic behavior. For a more elaborate distinction among relevant concepts and their many subtleties, see the book by Elliott Sober and David Sloan Wilson, *Unto Others: The Evolution and Psychology of Unselfish Behavior* (Cambridge, MA: Harvard University Press, 1999). For a warning label to this useful and comprehensive book, which is tinged by more than a hint of advocacy, see the review by J. Maynard Smith, "The Origin of Altruism," *Nature* 393 (1998): 639–640.

3. Perhaps the most comprehensive pertinent body of work is that of C. Daniel Batson, whose experiments up through the 1990s are summarized in Batson, *The Altruism Question: Toward a Social-Psychological Answer* (Hillsdale, NJ: Lawrence Erlbaum, 1991). A review of some of these experiments and why they are not

conclusive can be found in Sober and Wilson, *Unto Others*. Some argue that experiments of this sort can ultimately not decide the question whether people are sometimes driven by altruistic motives, see, e.g., Lise Wallach and Michael A. Wallach, "Why Altruism, Even Though It Exists, Cannot Be Demonstrated by Social Psychological Experiments," *Psychological Inquiry* 2 (1991): 153–155.

4. As usual, to adopt this perspective we have to give up something to gain something. What? You may not be able to accept this perspective at all. If so, why do you choose not to — for it is a choice you make?

5. Most of the behavioral examples I use here are standard fare that can be found in textbooks on animal behavior such as the excellent one by John Alcock, *Animal Behavior: An Evolutionary Approach,* 6th ed. (Sunderland, MA: Sinauer, 1998).

6. The evidence of how risky alarm calls are is mixed. Alcock, *Animal Behavior,* contains an elementary discussion.

7. U. Maschwitz and E. Maschwitz, "Platzende Arbeiterinnen: Eine neue Art der Feindabwehr bei Hautflüglern," *Oecologia* 14 (1974): 289–294.

8. The exceptions to this rule, such as primary and secondary helpers or brood parasitism in birds, are most instructive; see Alcock, *Animal Behavior*.

9. As seductively simple as the idea of fitness may be, some further warnings are in order. First, fitness emerges from a relationship among organisms or between an organism and the world around it. Use fitness with no such relationship in mind, and it becomes a hollow concept. Second and most important, the concept of fitness is a human construct. There may be no such thing as fitness "out there," and the concept sometimes fails to illuminate anything. Chance may sometimes be more important for survival or reproduction, and the possibility of "neutral evolution" resulting from random processes, and not driven by natural selection, has caused much debate among biologists. Problems surface, as they often do, when we forget that a concept like fitness is but a useful metaphor and reify it instead. This prevents us from abandoning the metaphor where useless or to modify it as needed. And any metaphor fails sometimes, in situations where the world is not as orderly as our concepts. One might think that we would take the occasional failure of a powerful metaphor like fitness as an opportunity, an opportunity for looking beyond it, beyond the fringes of our knowledge. Instead we tend to bury our heads in the sand in denial. Part of the reason stems from the concept of fitness itself. Precisely because the concept of fitness is so powerful and can explain so many observations, we delude ourselves into viewing it as universal, as being capable of explaining all features of living things.

10. Fitness and heritability are not necessarily related. Some features may be heritable but may not affect fitness. For example, my father's and my earlobes have similar shape, a shape I may have inherited. (I grew up separately from my father, and thus a shared environment cannot have shaped them similarly.) I could be mistaken, but the shape of a person's earlobes is not likely to affect fitness. Conversely, some features might not be heritable but affect fitness.

11. Luigi Luca Cavalli-Sforza and Marcus W. Feldman, *Cultural Transmission and Evolution: A Quantitative Approach* (Princeton, NJ: Princeton University Press, 1981).

12. If you know any chemistry, you might cringe here because I have not even mentioned charge, electronegativity, hydrogen bonds, or other properties of this molecule. Such properties are part of the language in which DNA conveys meaning, but the shape metaphor suffices for what I talk about here. I make this and even crasser simplifications for the sake of brevity. Consider that none of them affects the main message.

13. These are not only the reasons why inheritance is imperfect but also why there is inheritance to begin with. First, the environment does not determine an organism's features completely. Genes are important. Second, mistakes in copying genes are rare. And third, the offspring shares many thousand alleles with each parent—that is, the alleles packaged into the sperm and egg cells from which it originated.

14. Specifically, this blind spot is the abundance of cooperation, whether between genetically related organisms or genetically unrelated organisms. In line with the general observation that we tend to ignore blind spots, it took some one hundred years after *On the Origin of Species* was published for a more comprehensive viewpoint to be widely accepted. It is the viewpoint of inclusive fitness and kin selection; see W. D. Hamilton, "The Genetical Evolution of Social Behavior," *Journal of Theoretical Biology* 7 (1964): 1–16; and J. Maynard Smith, "Group Selection and Kin Selection," *Nature* 201 (1964): 1145–1147. Its germs can already be found in Darwin's *Origin of Species*.

15. Hamilton, "Genetical Evolution of Social Behavior."

16. One of the most forceful advocates and popularizers of this perspective is Richard Dawkins; see Dawkins, *The Selfish Gene,* 2nd ed. (Oxford: Oxford University Press, 1989). Perhaps the most authoritative, albeit more technical writing emphasizing one side of the dichotomy between self and other is that of E. O. Wilson; see, e.g., Wilson, *Sociobiology* (Cambridge, MA: Belknap Press of Harvard University Press, 1975). For a position distinctly opposed to an exclusively self-centered perspective on the living, see Frans De Waal, *Good Natured: The Origins of Right and Wrong in Humans and Other Animals* (Cambridge, MA: Harvard University Press, 1996); and Sober and Wilson, *Unto Others.*

17. This "selfish gene" perspective is advocated in Dawkins, *Selfish Gene.*

18. In doing so, two genetically related individuals need by no means be aware of their ulterior motives. They may feel utterly altruistic about their actions, but this fact itself, according to the selfish gene perspective, might be part of the devious strategy of their genes.

19. Dawkins, *Selfish Gene;* Alcock, *Animal Behavior;* Wilson, *Sociobiology.*

20. C. C. Mann, "Behavioral Genetics in Transition," *Science* 264 (1994): 1686–1689.

21. These events led to the evolution of multicellular organisms. Multicellularity

most probably evolved separately in algae, land plants, animals, and fungi. The example of *Volvox* suggests that its evolution may be much easier than we might think.

22. To make that simple a distinction is to slight the capabilities of bacteria for cooperation. For even they can live in elaborate associations, such as biofilms; see, e.g., H. Yasuda, "Bacterial Biofilms and Infectious Diseases," *Trends in Glycoscience and Glycotechnology* 8 (1996): 409–417; G. O'Toole, H. B. Kaplan, and R. Kolter, "Biofilm Formation as Microbial Development," *Annual Review of Microbiology* 54 (2000): 49–79. Biofilms are much like colonies with elaborate labor sharing and cooperation. But the individual cells forming such colonies can survive readily on their own, whereas cells in our body cannot.

23. Occasionally they may have sex, but that's a different story; see S. Baumberg et al., eds., *Population Genetics of Bacteria* (Cambridge: Cambridge University Press, 1995).

24. Tumors are examples of what happens if cells do not give up dividing.

25. For a different perspective on the evolution of multicellularity, revolving around organismal development, see Leo W. Buss, *The Evolution of Individuality* (Princeton, NJ: Princeton University Press, 1987).

26. A thorough introduction to this group of organisms is the monograph by David L. Kirk, *Volvox: Molecular Genetic Origins of Multicellularity and Cellular Differentiation* (Cambridge: Cambridge University Press, 1998). This group of organisms is fascinating for many reasons in addition to those listed below. For example, multicellularity may have originated as recently as thirty-five million years ago in this group, and may have evolved several times independently. There are also mutants of multicellular forms that generate unicellular forms. Moreover, some of the different multicellular forms can interbreed. All the evidence points to the relative ease with which multicellularity (and the self-sacrifice of individual cells) can evolve in this group of organisms. The existence of different species with different numbers of cells, and the observation that all these species make a successful living is significant also from an additional viewpoint. It tells us that each evolutionary step of adding a few cells might give some advantage to a multicellular organism and thus reduce the problem of explaining how organisms with billions of cells might have arisen.

27. Kirk, *Volvox*.

28. With the caveat that energy expenditure for transport in a viscous medium does not scale linearly with spatial extension.

29. With the exception of identical twins.

30. Are a gene and the organism harboring it also related in this sense?

31. For an elaboration of the example of wasps and figs, as well as for many other examples, see Friedrich G. Barth, *Insects and Flowers: The Biology of a Partnership* (Princeton, NJ: Princeton University Press, 1985).

32. The relationship between two symbionts is often not one-on-one. Especially if one partner spends only part of its life cycle with the other — as is the case

for insect pollinators or fruit-dispersing birds—this partner may have a very specialized lifestyle that might nevertheless involve various species. However, where the partners' life cycles are tightly interwoven, interactions are often exclusively one on one. For a much more nuanced discussion, with many examples, see John N. Thompson, *The Coevolutionary Process* (Chicago: University of Chicago Press, 1994).

33. Should nectar production be called "behavior"? For us, it will suffice to view behavior as any action serving a purpose.

34. That is, in analogy to organisms within a species, some of which are genetically related while others are not, one could distinguish organisms in different species that are related by fate from those that are not.

35. For examples, including those below, see Robert S. Desowitz, *New Guinea Tapeworms and Jewish Grandmothers: Tales of Parasites and People* (New York: W. W. Norton, 1981); and D. R. Linicome, "The Goodness of Parasites," in *Aspects of the Biology of Symbiosis,* ed. Thomas C. Cheng (Baltimore: University Park Press, 1971).

36. And what about the perfect predator, a veritable terminator wiping out all prey? In a similar vein, what would be the analogue of perfect selfishness, if self and other are organisms within the same species?

37. B. M. Sharp et al., "Origin and Evolution of AIDS Viruses," *Biological Bulletin* 196 (1999): 338–342.

38. J. J. Bull and I. J. Molineux, "Selection of Benevolence in a Host-Parasite System," *Evolution* 45 (1991): 875–882; S. L. Messenger, I. J. Molineux, and J. J. Bull, "Virulence Evolution in a Virus Obeys a Trade-Off," *Proceedings of the Royal Society B* 266 (1999): 397–404.

39. I am omitting many nuances of host-parasite interactions, but the general principles hold. Interestingly, one can also select for increased parasite virulence—parasite-inflicted host damage—by forcing parasite transmission through hosts genetically unrelated to previous hosts. For a departure point into some more technical literature, see J. J. Bull, "Virulence," *Evolution* 48 (1994): 1423–1437.

40. The now well-established theory of the origin of eukaryotic cells was originally proposed by Lynn Margulis; see Margulis, *Origin of Eukaryotic Cells: Evidence and Research Implications for a Theory of the Origin and Evolution of Microbial, Plant, and Animal Cells on the Precambrian Earth* (New Haven: Yale University Press, 1970). For a comprehensive recent review, see M. W. Gray, "The Endosymbiont Hypothesis Revisited," *International Review of Cytology* 141 (1992): 233–357.

41. I have not touched several interesting areas of intraorganismal conflict, including autoimmune diseases and intragenomic conflicts. In the latter case, maternal and paternal genes in one and the same organism may have conflicting "interests." Self and other may correspond to a mother and the embryo she harbors, which also carries paternal genes. Intragenomic conflict is intriguing, because the conflict arises within a single organism after the fertilization of an ovum. Even in such cases where self and other are physically very close, relatedness in fate,

and thus sameness, coexist with the antipodal difference of self and other that is reflected in their conflict.

42. I am purposely avoiding arguments from epidemiological theory. One such argument for sustained high parasite virulence is that if parasites get preferably transferred horizontally, between unrelated hosts, they have no "incentive" to evolve decreased virulence. Whether this argument holds up in finite host populations is currently unclear. Similarly, if multiple strains of a microparasite infect a host, then selection may increase virulence among them. In the case of the very same microparasites, however, group selection across host organisms is likely to lead to reduced virulence. In sum, current epidemiological thinking has not spoken the last word on host-parasite coevolution, especially considering the many complexities of how parasites and hosts can become linked in fate. Empirical evidence, however, makes the point that evolution of benevolence is by no means a general feature of coevolving species. For pertinent evidence, see Thompson, *Coevolutionary Process*.

43. The statements in this paragraph are expressed in anthropomorphic terms, but they could be translated into more precise neo-Darwinian language. For instance, in microparasites where multiple parasite strains can infect or coexist in the host, an evolutionary race among the strains may ensue in which the most rapidly reproducing strain outcompetes the other strains. This winning strain is often also the most detrimental to the host. Thus, the strain's short-term gain may be its long-term loss.

44. Very readable introductions with pointers to further literature on the prisoner's dilemma in human decision making and on the tit-for-tat strategy are Robert Axelrod, *The Evolution of Cooperation* (New York: Basic Books, 1984); and Axelrod, *The Complexity of Cooperation: Agent-based Models of Competition and Collaboration* (Princeton, NJ: Princeton University Press, 1997). For animal games, an older but still authoritative text is John Maynard Smith, *Evolution and the Theory of Games* (Cambridge: Cambridge University Press, 1982).

45. Axelrod, *Evolution of Cooperation*.

46. Elizabeth Roberts and Elias Amidon, eds., *Earth Prayers: From around the World: 365 Prayers, Poems, and Invocations for Honoring the Earth* (San Francisco: Harper, 1991). A systematic analysis of how traditional societies perceive relatedness among the living can be found in Fikret Berkes et al., "Exploring the Basic Ecological Unit: Ecosystem-like Concepts in Traditional Societies," *Ecosystems* 1 (1998): 409–415; and Enrique Salmon, "Kincentric Ecology: Indigenous Perceptions of the Human-Nature Relationship," *Ecological applications* 10 (2000): 1327–1332.

Chapter 3. Wholey Parts and Partly Wholes

1. There is of course a whole hierarchy of parts inside any cell, beginning with organelles and descending through proteins, lipids, and nucleic acids to ever smaller parts.

2. I am referring to the amino group and carboxyl group of proteinaceous amino acids.

3. The folding of a protein is a complicated process. We know that two identical chains of amino acids will form the same shape, meaning that the sequence of chain links itself determines a protein's final shape. Nevertheless, it is difficult to predict the final shape only from the sequence of amino acids in the chain.

4. Shape is by far not all that matters. Many other features are also important, including electric charge or electronegativity of amino acid residues. But the principle that I am after can be equally well illustrated with shape. Further details about this and other examples from this section can be found in textbooks of molecular biology and biochemistry, such as Lubert Stryer, *Biochemistry,* 4th ed. (New York: W. H. Freeman, 1995).

5. Numerous proteins are able to endure many amino acid changes in evolution, which apparently do not affect their function much, if at all. Such changes are called neutral changes, and they may have unexpected consequences later, perhaps millions of years later. Chapter 4 contains examples of such apparently neutral changes.

6. Ribozymes are not to be confused with ribosomes. To make proteins using the information contained in genes, a cell first transcribes DNA into messenger RNA, a molecule whose ribonucleotide building blocks are very similar to the deoxyribonucleotide building blocks of DNA. Ribosomes are complex molecular machines that use information in the ribonucleotide sequence of messenger RNAs to make the amino acid chains of proteins. Ribozymes, in contrast, are catalytic RNA molecules, RNA enzymes that are able to catalyze chemical reactions. (Ribosomes consist of many different protein and RNA molecules. Some of these RNA molecules are themselves ribozymes.) For a review on ribozymes, see Elizabeth A. Doherty and Jennifer A. Doudna, "Ribozyme Structures and Mechanisms," *Annual Review of Biochemistry* 69 (2000): 597–615.

7. This assumption has many manifestations in biology. It is reflected in the notion of keystone species (species that are critical to the integrity and composition of a community), and in the notion of master regulatory genes (genes that are central to the formation of organs or of any kind of macroscopic organismal feature). The latter kinds of genes are often fallaciously identified as genes "for" the feature, as if they were solely responsible for producing it. These notions are often taken to imply that some of the parts producing the whole are more important than others. However, this notion is rather simple-minded. For instance, population and quantitative genetic models reveal that nonlinear interactions among gene products can lead to a situation where changes in individual genes have disproportionately large effects on an organism but where the identity of these genes can change rapidly over evolutionary time, see e.g., Andreas Wagner, "Can Nonlinear Epigenetic Interactions Obscure Causal Relations between Genotype and Phenotype?" *Nonlinearity* 9 (1996): 607–629. Relatedly, repeated experiments that remove a keystone species from similarly structured communities show that the effects of such perturbations can vary widely. Moreover,

even parts that do not show dramatic effects on the whole when removed can be critical to the whole's stability in the face of other perturbations; see, e.g., Marie E. Csete and John C. Doyle, "Reverse Engineering of Biological Complexity," *Science* 295 (2002): 1664–1669. Stability of organismal features is as important to organisms as are the features themselves, and parts contributing to such stability may be as important as the more easily identifiable "keystone" or "master" parts.

8. Aside from being an intriguing phenomenon, isotope fractionation has proven useful to answer a variety of biological questions. Where does a plant obtain its minerals? What does an animal eat? From where did a population of butterflies migrate? Did prehistoric humans farm or hunt? Such questions can be addressed through a consequence of isotope fractionation, which is that the proportion of heavy atoms contained in an organism's tissues varies among species and among different environments. The reason is that different species harbor different enzymes that differ in their preference for light and heavy atoms. Different species thus differ also in the proportion of light and heavy atoms their bodies incorporate. If one is, for example, interested in an organism's food sources, one can take advantage of the fact that a dominant food source for the organism may leave its isotopic signature in the organism's tissues.

As a very speculative tangent, we could also imagine that A and B are themselves amino acid parts of proteins and that the proteins joining them are the kinds of proteins required to make proteins. (Proteins — more precisely, protein-RNA complexes called ribosomes — are critical for making proteins.) If a biologist provides to the cell only heavy amino acid A to make proteins, then all proteins made inside the cell would contain only heavy A. And just as some proteins prefer to use light A, because doing so takes less energy, they might fold differently if they contained heavy A, simply because it takes different amounts of energy to move amino acids of different mass. What, then, if some of the proteins that fold differently are proteins involved in making proteins?

9. Even where a change has no apparent immediate effect, such an effect can manifest itself in the future. We are immersed in illustrations of this principle through familiar wholes and their parts. Small changes in human societies, such as deteriorating living conditions, increasingly corrupt governments, and weakening economies, may change little for a long time. But at some point they may "tip" a society, whether through election, revolution, or war. The same holds for other wholes and their parts. Take proteins, where changes in a part may have no effects in the present but may instead affect its function in the future, perhaps millions of years hence and in conjunction with other change. Part of the reason is that a protein's function arises only as a result of a relation it has with the world of molecules around it.

10. For a more detailed account referring to much primary literature, see James T. Cushing, *Philosophical Concepts in Physics: The Historical Relation between Philosophy and Scientific Theories* (Cambridge: Cambridge University Press, 1998).

11. In this relation there is also space for the novel and surprising, that which drives a scientific conversation forward.

12. The following description draws heavily on the locomotion of bacteria like *Escherichia coli*. Other bacteria move in completely different ways, but flagellar locomotion has the merit of being unusually well studied. For a comprehensive introduction to existing work, see M. C. Macnab, "Flagella and Motility," and J. B. Stock and M. G. Surette, "Chemotaxis," both in *Escherichia Coli and Salmonella: Cellular and Molecular Biology,* ed. Frederick C. Neidhardt et al. (Washington, D.C.: ASM Press, 1996). A less technical review focusing on the biophysics of flagellar locomotion is provided by Howard C. Berg, "Motile Behavior of Bacteria," *Physics Today* 53 (2000): 24–29. In the account I provide here I simplify greatly to expose some principles that do not depend on details of the process under consideration.

13. Although "flagellum" derives from the Latin word for whip, it is actually misleading in the case of prokaryotic flagella, because these flagella are quite rigid. Eukaryotic flagella — like those of the Volvocaceae — whose structure is substantially different, are more whiplike in this sense.

14. The limitation of this analogy is that the smell of food in the air is usually not the food itself, whereas the substance drifting by the bacterium is.

15. Bacteria do not only swim toward some substances in the described way but also away from other substances, chemorepellents.

16. *From the perspective of the organism,* the decisions required from a bacterium and the most complex and difficult human decisions may be similar in that they have to be made in the face of limited information and foresight.

17. The name "cheA" derives from the abbreviation "che" for chemotaxis, the scientific name for the behavior of an organism that swims toward or away from particular chemicals. The letter "A" is an arbitrary suffix to this name.

18. Another component is necessary to understand the last link. A third enzyme, CheZ, which regenerates CheY from CheY-P, is constantly active in the cell. A cell will thus contain appreciable amounts of CheY-P only if CheA produces them. I am omitting other, similar details here that are not essential to the questions I am after. For these details, see Stock and Surette, "Chemotaxis," as well as articles specializing on the flagellar switch, e.g., Anat Bren and Michael Eisenbach, "How Signals Are Heard during Bacterial Chemotaxis: Protein-Protein Interactions in Sensory Signal Propagation," *Journal of Bacteriology* 182 (2000): 6865–6873; and Ruth E. Silversmith and Robert B. Bourret, "Throwing the Switch in Bacterial Chemotaxis," *Trends in Microbiology* 7 (1999): 16–22.

19. Notice that each of these causes is a *causa efficiens* in the Aristotelian sense, if one applies this terminology to bacterial locomotion.

20. One can use this simple system to illustrate the curious notion of a "gene for something." Imagine you are a scientist exploring how *E. coli* moves. You know very little about the molecules I talked about. But you can eliminate all of the CheA proteins in the cell or, better even, the one gene that contains the information to make CheA. What you would see is that the cell only swims straight. It never tumbles. Ah, you might say, this is the gene responsible *for* tumbling! Knowing what we know, this is of course an absurd conclusion. However, for many

other features of the living we are in a comparable situation, knowing very little about the parts that establish them. And we all have heard about genes *for* something, be it violent behavior, intelligence, homosexuality, or any other human feature.

21. How could one still make a policy choice if there was no one true answer about the origins of intelligence?

22. That both receptor and motor are complexes of many proteins further complicates the situation.

23. Why does a perspective focusing on one part ever work? And why is it so hugely successful in dissecting the architecture of the living? Because geneticists use an additional trick, not mentioned in the text. They work as much as possible with organisms that have the same genetic "background," and they allow change in only one molecular part of the organism. In other words, this perspective works if the *ceteris paribus* principle is applicable. In wild populations, of course, almost nothing is ever equal.

On a more technical note, one can make the case that perturbation of any gene (introduction of a different allele at a given gene locus) will always have an average affect on the phenotype when considered in the context of a population and that individual genes cause behavior in this precise sense and are selected for. This argument is perhaps as old as quantitative genetics itself but has been used most forcefully by George C. Williams, *Adaptation and Natural Selection: A Critique of Some Current Evolutionary Thought* (Princeton, NJ: Princeton University Press, 1966). The argument, however, works well only when individual genes contribute additively to a phenotype. If genes interact epistatically or nonadditively in the sense of quantitative genetics (as is usually the case), then the role of individual genes, and how strongly selection acts on them, can change substantially over time; see, e.g., Wagner, "Can Nonlinear Epigenetic Interactions Obscure Causal Relations between Genotype and Phenotype?"

24. But note that we needed to compare different bacteria to see it. The reason is that one and the same bacterium contains different forms — conformations — of each of the proteins involved in swimming, but it may contain only one and the same allele encoding each of these proteins.

25. "Slime molds" are a very heterogeneous group of species. This example is taken from slime molds in the genus *Dictyostelium*. For details, see John Tyler Bonner, *The Cellular Slime Molds* (Princeton, NJ: Princeton University Press, 1959); William F. Loomis, *Dictyostelium Discoideum: A Developmental System* (New York: Academic Press, 1975); and J. D. Gross, "Developmental Decisions in Dictyostelium discoideum," *Microbiological Reviews* 58 (1994): 330–351.

26. My apologies again to the expert for the omission of many important details in the interest of clarity. It was in Leo Buss's laboratory at Yale University that I had the opportunity to study the pumping polyps of *Hydractinia echinata* and *Podocoryne carnea*, two colonial hydrozoan species that inspired the examples used here. A more thorough introduction to hydrozoans than I can give here would be

provided by any textbook on invertebrate biology, e.g., Edward E. Ruppert and Robert D. Barnes, *Invertebrate Zoology,* 6th ed. (Fort Worth, TX: Harcourt College, 1994).

27. Brainless stomachs feeding each other may illustrate nature's wisdom. If so, then their pointless competition illustrates this wisdom's limitation, like a parasite's eradication of its host.

28. Because bacteria do not reproduce sexually like many other organisms, the notion of a bacterial species is built largely on distinct features of different bacteria, such as shape, colorimetric properties, or metabolic abilities. It is a typological species concept, as discussed in Douglas J. Futuyma, *Evolutionary Biology,* 3rd ed. (Sunderland, MA: Sinauer, 1998).

29. Gene transfer also occurs between organisms more akin to you and me, but it is exceedingly rare.

30. More precisely, one usually compares the DNA sequences of two separately evolving genes in such analyses and tries to estimate the time since their most recent common ancestor existed. Two separately evolving genes may arise through speciation or through gene duplications.

31. Not all of these clocks tick equally fast. Some genes may show many changes in a million years, others hardly any at all. Thus, some genes are useful to measure time on the scale of hundreds of thousands of years, whereas others are better suited to measure time in hundreds of millions of years. Wen-Hsiung Li, *Molecular Evolution* (Sunderland, MA: Sinauer, 1997), presents a thorough introduction to the topic.

32. Although this notion of a species does not apply to bacteria, the basic idea of the following passage, how to reconstruct evolutionary history, applies to both sexually and asexually reproducing organisms alike. (Asexually reproducing organisms are by definition reproductively isolated.)

33. Again, I have neglected that different molecular clocks tick at different rates. This problem can be alleviated by studying genes whose DNA sequence changes at comparable rates.

34. Howard Ochman, Jeffrey G. Lawrence, and Eduardo A. Groisman, "Lateral Gene Transfer and the Nature of Bacterial Innovation," *Nature* 405 (2000): 299–304; M.G. Lorenz and W. Wackernagel, "Bacterial Gene Transfer by Natural Genetic Transformation in the Environment," *Microbiological Reviews* 58 (1994): 563–602.

35. Some genes, especially those associated with information processing, seem to be transferred rarely between bacteria. One might use these genes to reconstruct history. But again, the question arises: the history of what?

36. I am leaving aside the well-known failure of many organisms to reproduce in captivity.

37. More information on the respective "species" *Ensatina escholtzii, Larus spp.,* and *Phylloscopus trochiloides* can be found in C. Moritz, C. J. Schneider, and D. B. Wake, "Evolutionary Relationships within the *Ensatina escholtzii* Complex

Confirm the Ring Species Interpretation," *Systematic Biology* 41 (1992): 273–291; and D. E. Irwin, S. Bensch, and T. D. Price, "Speciation in a Ring," *Nature* 409 (2001): 333–337.

38. As so often, physics provides an instructive anecdote. A few attributes suffice to describe an elementary particle. Some of these attributes are familiar (mass), yet others are exotic (including magnetic and orbital quantum numbers of electrons). Once having fixed these attributes, you can not say anything further about the particle. But if two elementary particles do have the same attributes, how are they different? They are not. They are one and the same particle, according to a principle of quantum physics known as the Pauli exclusion principle. It states that no two fermions (particles with spin one-half, including electrons, protons, and neutrons) can have the same attributes (energy quantum states). The many shades of gray in our everyday world, where two wholes may be distinguishable from one perspective but not from another, disappear here. Either two wholes are completely and clearly different or they are one and the same. Are these the ultimate parts, or are they wholes without parts?

Chapter 4. Risky Refuges

1. This statement about randomness needs some qualification. It is, for example, well established that changes from G or A to C and T are much less frequent than changes from G to A and from C to T. Similarly, certain nucleotide combinations, such as GC, show different propensity to suffer change than others. Compare these errors to the typographical errors you would make on a typewriter. They are not random either. The keyboard's layout and how you type (with how many fingers, etc.) will introduce statistical structure into your typographical errors. The point, however, is that these errors are random with respect to the meaning of the text. This points to the importance of choosing a proper reference point for randomness.

2. Based on an assumed mutation rate of 10^{-9} per replication per nucleotide and a gene length of five hundred nucleotides.

3. Observing more DNA changes in a starving colony does not imply that all cells in it increase their mutation rate. However, the cells that increase their mutation rate may be most likely to break out of starvation.

4. One could simply say that starving cells shut down DNA repair, but that would be imprecise. For example, under many stresses, including that of excessive DNA damage or of nutrient deprivation, bacterial cells activate the so-called SOS response. Some of the genes turned on during this SOS response produce error-prone DNA polymerases involved in DNA repair. The result is an elevated frequency of mutations.

That cells actively turn on genes required to deal with stress also makes it unlikely that cells simply run out of energy to keep repair functions up. Patricia L. Foster, "Adaptive Mutation: Implications for Evolution," *Bioessays* 22 (2000):

1067–1074, provides a more detailed discussion of the increased number of mutations in nutritionally deprived cells. The fact that the SOS response involves the activation of several DNA polymerases, only some of which have higher mutation rates, makes it seem unlikely that higher mutation rates are an inevitable response to nutritional deprivation and that nothing about them reflects an evolutionary adaptation.

It is important to distinguish changing mutation frequencies in response to stress from the evolution of mutation rates over many generations, an evolutionary process I do not explicitly address in this section. See, e.g., Paul D. Sniegowski et al., "The Evolution of Mutation Rates: Separating Causes from Consequences," *Bioessays* 22 (2000): 1057–1066.

5. In a starving colony, would most cells die from starvation or from mutations wreaking havoc on them? The answer, it appears, is starvation; see, e.g., Foster, "Adaptive Mutation." Here and below, I use anthropomorphic language for convenience. It should be understood that the processes at issue, such as regulatory processes causing expression changes in DNA repair proteins, are shaped by natural selection.

6. Most of the experimentally well studied examples do actually not involve surface changes in existing proteins, but—for ease of experimental analysis— mutations that lead to the excision of mobile genetic elements from genes of genetically manipulated strains. Such excision eliminates DNA that "silenced" these genes, allowing the expression of hitherto unexpressed proteins.

7. In the late 1980s, experiments by John Cairns and collaborators suggested that nutritionally deprived cells might be able to bias occurring mutations towards those beneficial to the cell. This phenomenon of "directed mutation," however, has not survived scrutiny. P. D. Sniegowski and R. E. Lenski, "Mutation and Adaptation: The Directed Mutation Controversy in Evolutionary Perspective," *Annual Review of Ecology and Systematics* 26 (1995): 553–578, contains an overview of relevant material.

8. The limitation of the analogy is that the odds to arrive at an improved combustion engine are probably much smaller than the odds of creating a bacterium capable of using a new food source. The reason may be that organisms have evolved to be evolvable through this gamble.

9. This statement is based on observations on the rate of excision of the bacteriophage Mu from a location between the *ara* and *lac* operons in the bacterium *E. coli*. These excisions allow the bacterium to grow on lactose as a carbon source. They occur at a frequency of (much) less than 10^{-9} per cell and day under nonstarvation conditions, but at a frequency of 10^{-5} in starving cells, (much) more than ten thousand times higher. See, e.g., R. E. Lenski and J. E. Mittler, "The Directed Mutation Controversy and Neo-Darwinism," *Science* 259 (1993): 188–194. The work of Barry G. Hall contains other instructive cases, e.g., in Hall, "Activation of the Bgl Operon by Adaptive Mutation," *Molecular Biology and Evolution* 15 (1998): 1–5.

10. A word about viewing bacteria as decision makers is in order here. If such language is disturbing, it may help to recall our earlier discussion of parts and wholes, human or otherwise, where I mentioned the proteins detecting and influencing a bacterium's swimming direction. It was a matter of choice then where to place responsibility for any change in swimming direction, on the whole bacterium, on a few parts, or even one part of the whole. This perspective also applies to starving bacteria, because each cell harbors proteins that detect starvation, similar to those detecting swimming direction. These proteins then influence others that change a cell's DNA repair crew, a change that may lead a cell out of starvation. The agent of change is a matter of your choice and perspective. Notice that we tacitly apply the same principle to humans. In calling something a human decision, we choose a perspective centering on the human whole. (For us, this perspective is especially attractive because we understand only dimly the role our brain and molecular body parts play in our "decisions.")

11. Here is an interesting although perhaps entirely superficial analogy. A large body of work on the psychology of choice suggests that humans, when faced with (catastrophic) loss, tend to accept gambles to avoid that loss. They become risk seekers. In contrast, when faced with a choice between a certain small gain and an uncertain larger gain, humans act risk-aversely. They tend to accept the certain smaller gain; see, e.g., A. Tversky and D. Kahnemann, "The Framing of Decisions and the Psychology of Choice," *Science* 211 (1981): 453–458.

12. The importance of values is well recognized by analysts of risk in human decisions; see Baruch Fischhoff, Stephen R. Watson, and Chris Hope, "Defining Risk," *Policy Sciences* 17 (1984): 123–129. For a humorous perspective, see Mark Twain's essay "The Danger of Lying in Bed," from which the chapter's epigraph is taken, in S. L. Clemens, *The Unabridged Mark Twain* (Philadelphia: Running Press, 1976).

Among a wide spectrum of perspectives about risk in the world, two extremes deserve special mention. According to one, the world is full of risks and devoid of safety, a terrifyingly dangerous place. According to its opposite, nothing is risky, the world is absolutely safe. Although you might agree that life is terribly dangerous, the opposite might seem absurd at first sight. For dangers lurk everywhere: in electric outlets (five hundred people are electrocuted annually in the United States); in foods as pedestrian as peanut butter (improperly stored peanuts contain aflatoxin, a potent fungal carcinogen); in the air (twenty thousand people die annually from air pollution in the eastern United States); in the mountains (high altitude increases cancer risk due to cosmic radiation); in chlorinated drinking water (it forms potent toxins such as chloroform); in flammable pajamas; or in a distracted driver crossing my path as I drop this manuscript off at the post office. Many such risks are analyzed more thoroughly in Richard Wilson, "Analyzing the Daily Risks of Life," *Technology Review* 81 (1979): 41–46; and M. Granger Morgan, "Probing the Question of Technology-Induced Risk," *IEEE Spectrum* 18 (1981): 58–64. In light of these observations, most of us would ask how the world could *possibly* be an absolutely safe place? Yet some choose this perspective. Just take the famous Persian sage Rumi, who said (to who knows whom), "*I trust you to kill me*" (Coleman Barks, trans., *The Essential Rumi* [Edison, NJ: Castle Books, 1997]). If nothing else, this statement underlines that safety and risk involve choice, as well as a perspective built on choice.

13. For an immensely readable popular introduction with many technical references to the topic of risk and uncertainty and their history in human society, see Peter L. Bernstein, *Against the Gods: The Remarkable Story of Risk* (New York: Wiley and Sons, 1996).

14. The distinction originated in Frank H. Knight, *Risk, Uncertainty and Profit* (1921; reprint ed., New York: Century, 1964).

15. K. J. Arrow, "I Know a Hawk from a Handsaw," in *Eminent Economists: Their Life and Philosophies,* ed. M. Szenberg (Cambridge: Cambridge University Press, 1992) 42–50.

16. Bernstein, *Against the Gods.*

17. This is the reason why decision theory in psychology and economics uses only games with well-defined, usually monetary outcomes. The decisions in such games are easy to analyze quantitatively, but they constitute a tiny segment of all decisions made in human life.

18. And why is this just a perspective? Because one day we might have not only complete information but also agreement on all our values. But what are the chances of *that*? And would you care to live in such a world?

19. Here is a simple experiment to convince yourself of the veracity of this statement. Put one — only one — food source, say glucose, in a laboratory culture of bacteria. When running out of food, the bacteria in the culture will starve and eventually die. This is how serendipity, a freak accident of history — your action — could end the three-billion-year-long winning streak of the cells in this culture.

20. Mutation frequencies can be defined in a variety of ways, as mutations per generation, per round of DNA replication, or per year. Moreover, one can count the number of mutations per genome, per gene, or per nucleotide. To compare these different kinds of mutation frequencies is not straightforward.

21. The inverse relationship between mutation rate and genome size for microbes is discussed in John W. Drake et al., "Rates of Spontaneous Mutation," *Genetics* 148 (1998): 1667–1686. For unicellular and multicellular organisms taken together, the relationship between mutation rate and genome size may become positive, which may be partly caused by the reduced efficacy of selection in higher organisms with small population sizes, e.g., Michael Lynch, *The Origins of Genome Architecture* (Sunderland, MA: Sinauer, 2007).

22. This interpretation may provide a coarse explanation for a general pattern. However, it is likely to be insufficient in explaining every aspect of this pattern. For example, DNA content and gene content do not show a one-to-one correspondence. Also, there is a cost to repairing DNA mutations, and this cost increases the more mutations an organism eliminates. At some point, this cost may be-

come prohibitively high, so that mutation frequencies cannot fall below certain thresholds.

23. Another case in point is the high frequency, up to 20 percent, of mutator strains (strains with elevated mutation frequencies) observed in wild isolates of various bacterial species, as well as in laboratory evolution experiments, e.g., Sniegowski et al., "Evolution of Mutation Rates."

24. Notice that this is not an explanation of how this ability came about in the first place. Such an explanation is much more difficult to come by. Evolution takes place when individuals in a population have different and heritable abilities to survive and reproduce. The ability of any one bacterium to shut down DNA repair will probably *reduce* its ability to survive and reproduce. This is because the bacterium itself is not likely to benefit from its gamble. Its behavior must benefit a bigger whole, which is then better able to survive and reproduce. This whole may not be the individual bacterium but, for example, the colony. The question whether natural selection can act on wholes different from the individual has been a matter of much debate.

25. As a cautionary note, too many factors associated with this process are unknown to be certain that it is of universal importance. For example, its success depends on beneficial mutations being sufficiently frequent, recombination being sufficiently rare, and the cost of repair being similar across organisms. Some experts think that the observed mutation rates reflect, in most cases, the cost of repair rather than a gamble, e.g., Sniegowski et al., "Evolution of Mutation Rates." It is also important to note that organisms do not anticipate the need to gamble for their future benefit. Rather, if they gamble, it is because gambling was successful in the past.

26. Or at least we cannot find a way in which they do. Notice that I do not use the notion of neutrality in the narrow sense of a genetic change that does not affect fitness, as in Motoo Kimura, *The Neutral Theory of Molecular Evolution* (Cambridge: Cambridge University Press, 1983). Instead, I call a neutral change a change that does not affect one aspect of an organism's performance in its present environment.

27. Possible exceptions include variation-generating hypermutation mechanism, for example in the generation of antibodies, where the creation of diversity (and not of a specific variant) itself is advantageous.

28. I will here focus on neutral losses of genes, proteins, and lifestyle features. But the same principle applies to neutral modifications of existing proteins, modifications that can become important later on. An excellent example includes the evolution of lactalbumin, a protein essential for production of lactose in mammals, from an enzyme that served as an antibiotic, killing bacteria, e.g., Richard E. Dickerson and Irving Geis, *The Structure and Action of Proteins* (New York: Harper and Row, 1969).

In general, whether novel organismal features arose through changes neutral with respect to *every* aspect of the organism's life or whether all of the required changes served some purpose — whether they arose as exaptations or nonaptations in Stephen Jay Gould's and Elisabeth S. Vrba's diction (Gould and Vrba, "Exaptation: A Missing Term in the Science of Form," *Paleobiology* 8 [1982]: 4–15) — has been a bone of much contention. For my purpose, this controversy is peripheral, because the point is that changes neutral *with respect to a particular lifestyle* feature can have unexpected consequences.

29. Some organisms know no "vitamins," others know fewer vitamins, and some know more than we do.

30. Russell F. Doolittle, "Microbial Genomes Multiply," *Nature* 416 (2002): 697–700.

31. These resistance mechanisms can be as varied as the drugs themselves: they may protect the parts of a bacterial cell attacked by the drug, pump the drug out of the cell, or disarm the drug chemically.

32. Some of these purgings may in fact be beneficial, which should not distract from the main argument, that the consequences of change cannot be foreseen except in our toy models of the world.

33. Animals in general cannot synthesize vitamin A (retinol) *de novo*.

34. A neutral feature is conventionally viewed as a feature that does not provide any benefit to an organism. Recall that I use the notion of neutrality in a different sense, namely as a feature neutral with respect to a particular aspect of an organism's lifestyle, such as flying.

35. This and other examples of morphological innovations are explored from a molecular perspective in the following book, which also contains further references: Sean B. Carroll, Jennifer K. Grenier, and Scott D. Weatherbee, *From DNA to Diversity: Molecular Genetics and the Evolution of Animal Design* (Malden, MA: Blackwell, 2001).

36. Some extreme purgings that I talked about above — although I called them neutral for brevity — also illustrate this point. Such purgings may not be merely neutral but even beneficial, freeing up abundant space in the "garage." For instance, losing a thousand genes may free up energy not spent in copying DNA, repairing it, and producing proteins. However, as I said earlier, purgings can quickly turn fatal if a lost feature becomes necessary. The question whether loss of unnecessary features is neutral or beneficial mirrors exactly the tension between neutral and beneficial changes in the evolution of new functions of organismal parts, including proteins. Small losses may be neutral and very large losses are beneficial, but where to draw the line is much less clear.

37. The opposite orientation, seven (paired?) tentacular legs and fourteen dorsal spines, has also been proposed, e.g., in L. Ramsköld and Hou Xianguang, "New Early Cambrian Animal and Onychophoran Affinities of Enigmatic Metazoans," *Nature* 351 (1991): 225–228. We may never have certainty about the truth.

38. A popular introduction to this earliest fauna is Stephen Jay Gould, *Wonderful Life: The Burgess Shale and Nature of History* (New York: W. W. Norton, 1989).

39. These examples and many others can be found at the Invasive Species Specialist Group (www.issg.org).

40. Another possible cause is that constraints on the development and evolution

of such organisms prevent change. Popular introductions to the subject of living fossils are Niles Eldredge, "Survivors from the Good Old, Old, Old Days," *Natural History* 81 (1975): 60–69; and Peter Douglas Ward, *On Methuselah's Trail: Living Fossils and the Great Extinctions* (New York: W. H. Freeman, 1992). For the many technical subtleties in identifying living fossils, see Niles Eldredge and Steven M. Stanley, eds., *Living Fossils* (New York: Springer, 1984).

41. The Greek word "telos" means end or purpose. Ernst Mayr, "The Idea of Teleology," *Journal of the History of Ideas* 53 (1992): 117–135, distinguishes cosmic teleology from other, less problematic kinds of teleological (end-directed) phenomena.

42. Perhaps the best-known exponent of this perspective on biological evolution is Pierre Teilhard de Chardin; e.g., Teilhard de Chardin, *Christianity and Evolution*, trans. René Hague (New York: Harcourt Brace Jovanovich, 1971).

43. At Darwin's time, people also considered another possibility. According to this "Lamarckist" hypothesis, giraffe's ancestors stretched their necks to reach up higher into trees. Over a lifetime, their necks would grow longer through such exercise. This acquired long neck would then be inherited to the next generation. Over many generations, giraffes acquired longer and longer necks by inheritance of well-exercised stretched necks. The problem is that stretched necks acquired by exercise are not inherited, just as strong muscles themselves are not. (Although the ability for neck stretching might be, but that's another story.)

44. A more elaborate exposition of giraffe biology can be found in Anne Innis Dagg and J. Bristol Foster, *The Giraffe: Its Biology, Behavior, and Ecology* (New York: Van Nostrand Reinhold, 1976).

45. D. W. Shea, "Complexity and Evolution: What Everybody Knows," *Biology and Philosophy* 6 (1991): 303–324, provides a more differentiated discussion that summarizes earlier efforts to identify trends toward increased complexity. The question whether life shows any trends — related to complexity or otherwise — is discussed in M. Ruse, "Evolution and Progress," *Trends in Ecology and Evolution* 8 (1993): 55–59.

46. If nobody has come up with widely accepted and still nontrivial measures of complexity, it has not been for lack of trying. A growing community of scientists — scientists who study human behavior, swimming bacteria, gyrating stock markets, and nest-building ants — is interested in what has come to be known as the science of complex systems. The problem is that except for narrow areas such as computer science, there is no widely agreed upon operationally useful (!) criterion to distinguish a simple from a complex system. Criteria stemming from physics, such as entropy or information content, have not been helpful, since it is unclear how to measure them in practice, for example, to assess the complexity of a living cell.

47. To see this as a possibility, you may want to revisit an earlier part of our conversation. Can one thing be more complex than another if part and whole — or part of part of part, and whole — are two sides of the same coin? (What prevents us from viewing bacteria as equally complex to humans?)

48. Two introductory expositions to the history of computing are Georges Ifrah, *The Universal History of Computing: From the Abacus to the Quantum Computer*, trans. E. F. Harding (New York: Wiley, 2001); and Michael R. Williams, *A History of Computing Technology* (Los Alamitos, CA: IEEE Computer Society Press, 1997).

49. The fascinating topic of laboratory evolution, which may also provide an inroad into this problem, deserves mention here. The evolution of small organisms with short generation times, such as fruit flies or bacteria, can be easily studied over many generations in the laboratory. Part of the reason is that organisms like these have short generation times and that they can sustain large population numbers (in the thousands for fruit flies, in the millions for bacteria) in a small space. A representative example is the work of Richard Lenski and collaborators, who carried out independent and parallel evolution experiments in bacteria over tens of thousands of generations, starting with genetically identical clones. Each parallel evolution experiment was carried out in the same environment. See, e.g., Richard E. Lenski and Michael Travisano, "Dynamics of Adaptation and Diversification: A 10,000-Generation Experiment with Bacterial Populations," *Proceedings of the National Academy of Sciences* 91 (1994): 6806–6814. The growth rate (fitness, in these populations) of each evolved lineage increased relative to the starting clone, as did the cell size. It is intriguing that these parameters did not increase to the same extent in parallel cell lines, nor was the change in cell size always correlated with the increase in fitness across lines. Different cell lines may have arrived at higher fitness through varying avenues (different mutations) although detailed analysis of the mutations in question is still under way. Whatever these mutations are, it is clear that life can and does turn out differently in each test tube.

50. Much finer scale biogeographic comparison reveals that convergent or parallel evolution occurred in many lineages that have otherwise unique features. For example, although marsupials are endemic to Australia, many of their adaptations reflect those found in organisms of other continents. On a smaller scale, anoles lizards are found in convergently evolved ecomorphs on the islands of Hispaniola and Puerto Rico. That convergent evolution is not universal illustrates that biological evolution is no exception to the closeness of unpredictability and lawfulness. A paleontological perspective on the importance of history in biological evolution is taken in Gould, *Wonderful Life*.

51. The revocation of commitment here is often not caused by any inferiority of the adopted product or service. This phenomenon is closely related to the development and change of technology standards mentioned in the context of computing technology. Stanley M. Besen and Joseph Farrell, "Choosing How to Compete: Strategies and Tactics in Standardization," *Journal of Economic Perspectives* 8 (1994): 117–131; and R. Cowan, "High Technology and the Economics of Standardization," in *New Technology at the Outset,* ed. M. Indardiza and U. Hoffman (Frankfurt: Campus, 1992), provide illustrative examples.

Chapter 5. Destructive Creation

1. However, it is of more recent provenance in biology, having been coined only in 1972; see J. F. Kerr, A. H. Wyllie, and A. R. Currie, "Apoptosis: A Basic Biological Phenomenon with Wide-Ranging Implications in Tissue Kinetics," *British Journal of Cancer* 26 (1999): 239–257.

2. To be precise, there are two major kinds of plant vascular tissue: phloem and xylem. Xylem transports water from the roots to all plant parts. Phloem transports nutrients throughout the plant. The role of programmed cell death has received significant attention in the formation of xylem; see, e.g., Zheng-Hua Ye, "Vascular Tissue Differentiation and Pattern Formation in Plants," *Annual Review of Plant Biology* 53 (2002): 183–202. As opposed to animals, where apoptosis generally results in complete cell elimination, this is often not the case in plants, where a rigid cell wall would prevent such elimination. R. I. Pennell and C. Lamb, "Programmed Cell Death in Plants," *Plant Cell* 9 (1997): 1157; and Eric Lam, "Controlled Cell Death, Plant Survival and Development," *Nature Reviews Molecular Cell Biology* 5 (2004): 305–315, review the role of apoptosis in plants.

3. Stephen M. Stahl, *Essential Psychopharmacology: Neuroscientific Basis and Practical Applications,* 2nd ed. (New York: Cambridge University Press, 2000), 24.

4. Lewis Thomas, *The Fragile Species* (New York: MacMillan, 1992).

5. The actual number may be substantially higher. For instance, Robert C. Bast et al., *Cancer Medicine 5,* 5th ed. (Hamilton, ON: B. C. Decker, 2000), section 1.2, cites evidence that fifty to seventy million cells perish apoptotically each day in an average adult. At this rate, an individual replaces cells amounting to his or her entire body mass every year.

6. For some more examples and many more details about apoptosis in development, see Pascal Meier, Andrew Finch, and Gerard Evan, "Apoptosis in Development," *Nature* 407 (2000): 796–801.

7. Some cells arrive at the decision to die on their own. In the course of their life, they have suffered excessive and irreparable damage. Perhaps their DNA has been broken into pieces through radiation, or a virus has wrested control over their lives, and turned them into factories that make new virus particles. Many such cells initiate a communication process among their parts, a process whose outcome is death. This process ultimately also leads to cell death through activation of proteases that lay waste to the cell's parts.

8. For a glimpse at many of the molecular details that I am omitting here, see Michael O. Hengartner, "The Biochemistry of Apoptosis," *Nature* 407 (2000): 770–776.

9. As usual, this statement is a simplification, because molecular forces such as electrostatic attraction and repulsion also play a role.

10. One important detail that I have omitted regards the role of positive signals to keep cells alive. That is, in addition to molecular conversations that can kill a cell, others are needed to keep cells alive. Examples include the importance of

neurotrophic factors secreted by neurons, which are necessary to keep other neurons alive, and the importance of a cell's attachment to the extracellular matrix surrounding it. Once detached from this matrix, many cells undergo apoptosis. The abundance of such positive signals has led to the suggestion that cell death is the default for animal cells and that cells need to be kept alive through positive signals; see, e.g., Martin C. Raff, "Social Controls on Cell Survival and Cell Death," *Nature* 356 (1992): 397–400.

11. If this seems a strange perspective on suicide, ask yourself whether human suicide is fundamentally different. That is, do messages sent by others play no role in human suicide?

12. Biochemists are uncovering an increasing number of proteins that cleave themselves into pieces. Such autoproteolytic proteins, however, do not only self-destruct. They may also self-create: the cleaved form may fold into a new shape, allowing it to catalyze new chemical reactions.

13. Lubert Stryer, *Biochemistry,* 4th ed. (New York: W. H. Freeman, 1995), 942–945, contains further background information on protein degradation. The listed amino acids confer differential stability on proteins in yeast. The list is not complete, nor do the same amino acids confer (in)stability in all species.

14. As quoted in Timothy Ferris, *The Whole Shebang: A State-of-the-Universe(s) Report* (New York: Simon and Schuster, 1997).

15. Michael R. Rose and Caleb E. Finch, eds., *Genetics and Evolution of Aging* (Dordrecht: Kluwer Academic, 1994).

16. Strictly speaking, it depends on the metabolic rate, or the amount of energy consumed per gram of body weight. Metabolic rate — and thus longevity — also scale with body mass; see James H. Brown, Geoffrey B. West, and Brian J. Enquist, "Scaling in Biology: Patterns and Processes, Causes and Consequences," in *Scaling in Biology,* ed. Brown and West (Oxford: Oxford University Press, 2000). Metabolic rate is lower for larger organisms, which thus live longer than smaller organisms. This may also be the reason why mild starvation, technically known as caloric restriction, is the only known effective means of prolonging an organism's life span. Rats under lifelong caloric restriction can show an up to 50 percent increase in life span; see, e.g., A. Richardson and M. A. Pahlavani, "Thoughts on the Evolutionary Basis of Dietary Restriction," in *Genetics and Evolution of Aging,* ed. Rose and Finch.

17. Genes associated with longevity span a range of functions, from genes encoding superoxide dismutase, an enzyme needed for the cleanup of free radicals, to genes encoding heat shock proteins, which protect other proteins against thermal denaturation, and genes encoding proteins involved in bodily defenses against diseases. This small sample from a broad spectrum of gene functions indicates that we cannot hope for a universal answer to the question that genes drive aging. For a review, see James W. Curtsinger et al., "Genetic Variation and Aging," *Annual Review of Genetics* 29 (1995): 553–575. In some species it has also been shown that

aging involves many genes and shows low heritability, indicating strong environmental influences.

18. According to prevailing opinion, the ultimate or evolutionary cause of aging is that the reproductive potential of most organisms is small late in life when compared to early in life. Therefore, only weak or no selection takes place against genes that cause senescence, as long as such senescence arises late in life. Two main evolutionary hypotheses rest on this principle. One proposes a mechanism of antagonistic pleiotropy or a trade-off between fitness (reproduction and viability) early in life and late in life. In other words, genes that increase fitness early in life decrease fitness late in life. The other hypothesis posits that because of the decreasing effectiveness of selection late in life against genes causing senescence, mutations increasing mortality late in life are free to accumulate over time. Both hypotheses have empirical support. For more comprehensive treatments of the genetics and evolution of aging, see Michael R. Rose, *Evolutionary Biology of Aging* (New York: Oxford University Press, 1991); Brian Charlesworth, *Evolution in Age-Structured Populations* (Cambridge: Cambridge University Press, 1980); Curtsinger and Fukui, "Genetic Variation and Aging"; and an older but seminal article by George C. Williams, "Pleiotropy, Natural Selection, and the Evolution of Senescence," *Evolution* 11 (1957): 398–411. Notice that nothing in the above arguments argues that aging and death serve a purpose to a population. It is easy to construe such arguments, but they would rely on group or species selection. Whether such group or species selection is frequent in higher organisms is controversial.

19. Max K. Planck, *Scientific Autobiography and Other Papers* (New York: Greenwood, 1968).

Chapter 6. Choice in the Fabric of Chance and Necessity

1. Some laws of nature, such as quantum mechanical laws, prevent certain kinds of predictions. However, in addition to their explanatory power, such laws often also increase predictive power in other areas.

2. According to the classical view, as you heat a solid, its constituents vibrate more and more vigorously in their rigid positions, until they break free and form a liquid or gas. The nature of heat raises intriguing questions in the philosophy and history of science, some of which are addressed in Stathis Psillos, "A Philosophical Study of the Transition from the Caloric Theory of Heat to Thermodynamics: Resisting the Pessimistic Meta-Induction," *Studies in the History and Philosophy of Science Part A* 25 (1994): 159–190.

3. In other words, the ink molecules diffuse through the water, a process during which the disorder (entropy) of the water-ink system increases. The initial state, in which all ink molecules are concentrated in one ink droplet, is a highly unlikely, low-entropy, far-from-equilibrium state that gives rise to a much more likely, high-entropy, equilibrium configuration in which ink molecules are spa-

tially nearly uniformly distributed in the liquid and in which our uncertainty about their position is maximal.

4. Chemistry textbooks are full of examples, such as the chlorination of many alkanes, which yield more than one product.

5. An alternative formulation is that the formation of x percent C and y percent D is the law of nature of interest. However, the uncertainty does not go away for this formulation either, because the respective percentages always come with a margin of error.

6. Sometimes the outcome is not predictable, as in the case of individual choice, thus yielding an example of unpredictability emerging from the unpredictable. I choose to focus here on the predictable that emerges from the unpredictable and the unpredictable that emerges from the predictable, simply because they are, at first sight, not as easy to understand as their two counterparts, the predictable emerging from the predictable and the unpredictable emerging from the unpredictable.

7. Thomas C. Schelling, "Dynamic Models of Segregation," *Journal of Mathematical Sociology* 1 (1971): 143–186.

8. The matter is usually much more complex than in the caricature I present here. Many other factors play a role, such as the population density of host and disease agents, dispersal properties of the host population, nonrandom patterns of host interactions, evolution within the parasite populations, especially through host-superinfection, and so on. A technical review of relevant work is given in Roy M. Anderson and Robert M. May, *Infectious Diseases of Humans: Dynamics and Control* (New York: Oxford University Press, 1991), who also estimate that the critical population coverage of immunization necessary to prevent disease lies between 70 percent and 99 percent, depending on the disease.

9. The basis of all this is the well-worn observation that all laws of nature are laws of probability, although this is more easily recognized in some areas, say for laws in the social sciences, than in others, such as the laws of classical mechanics.

10. To be precise, there are stable periodic solutions to many chaotic systems, often beautifully complex ones, and some mathematicians have turned the search for such solutions into a sport.

11. The point is sometimes made that predictability and determinism are not the same. It may seem that I do not distinguish between them. That is true, and here is the reason. To be sure, we can think of trivial kinds of unpredictability that exist despite determinism, exemplified by a clock that is locked up in a safe that nobody can access. The clock will proceed to work — deterministically — although we cannot know its state. (The example is taken from J. Bricmont, "Science of Chaos or Chaos in Science," in *The Flight from Science and Reason,* ed. Paul R. Gross, Norman Levitt, and Martin W. Lewis [Baltimore: Johns Hopkins University Press, 1996].) But I am much more concerned with unpredictability of a different, nontrivial kind, where neither we *nor any other finite mind* could predict, in principle or in practice, the future of a system. I maintain that unpredictability

of this nature is the same as indeterminism. Fundamentally, this proposed equivalence rests on the assumption that the mathematical laws from which deterministic chaos emerges and the physical laws that limit predictability are universal and that any finite intellect is subject to them. The contrary opinion emphasizing the "*in principle*" predictability that a Laplacian spirit is capable of is expressed in Bricmont, "Science of Chaos or Chaos in Science." In my view, the fallacy behind the latter view is similar to that of viewing a mental construct like infinitesimal calculus as more than what it is — a product of our imagination.

12. Here chaos simply stands for unpredictable behavior. It is deterministic, because it emerges from a law telling you everything that matters about the bodies' motion. Technically speaking, deterministic chaos is characterized as sensitivity of a dynamical system to initial conditions, which ultimately implies unpredictability given measurements of finite precision. A popular introduction to chaotic dynamics is James Gleick, *Chaos: Making a New Science* (New York: Viking Penguin, 1987), and a more technical one is Steven H. Strogatz, *Nonlinear Dynamics and Chaos* (Reading, MA: Addison Wesley, 1994).

13. Although many laws are expressed in terms of differential equations, discrete difference equations such as the logistic map describing population growth of a species with discrete generations are notable exceptions. However, the infinitesimal enters these maps as well, in the sense that very small differences in initial conditions can be very important here also.

14. Again, unless you admit — a game of pretense — an infinite mind, unpredictability is the same as indeterminism. How the most complex metaphors of mathematics — including that of infinity — emerge from our everyday pedestrian thinking is beautifully illustrated in George Lakoff and Rafael E. Núñez, *Where Mathematics Comes From: How the Embodied Mind Brings Mathematics into Being* (New York: Basic Books, 2000).

15. The most famous such example is the Belousov-Zhabotinskii reaction involving the reaction of organic material with bromate in an acidic environment. See, e.g., J. D Murray, *Mathematical Biology* (New York: Springer, 1993).

16. On a historical note, deterministic chaos in organismal populations was not first uncovered in models of interactions of two or more species, although it has gained great prominence in this area, for example in models of chaotic epidemics or in the lynx-hare dynamics. It has first been described in detail for the chaotic dynamics of a logistically growing population of a *single* species with discrete generations; see Robert M. May and George F. Oster, "Bifurcations and Dynamic Complexity in Simple Ecological Models," *American Naturalist* 110 (1976): 573–599. An important difference of ecological systems to other systems such as chemical reaction networks is that the empirical data to demonstrate chaotic dynamics is very hard to come by in ecological systems. Thus, many proposed cases of erratic behavior in ecological communities may be due to sources of stochasticity other than deterministic chaos.

17. Lawfulness and unpredictability also exist only in relation to each other,

because the lawful is a departure from the irregular and the irregular can be seen only in reference to law. Could one even envision a world completely devoid of laws and regularities? Conversely, what would a completely predictable world look like? Could there be life in such a world?

18. I should mention where this position on choice is situated in relation to the free will debate, which deals with the narrower question whether humans have choices. (You can find a concise summary of key positions in this debate in a collection of articles edited by Joel Feinberg, *Reason and Responsibility: Readings in Some Basic Problems of Philosophy,* 5th ed. [Belmont, CA: Wadsworth, 1981].) As might be obvious by now, I do not take a deterministic perspective, at least not in the superficial sense that the future, near or far, is knowable. It may seem that the perspective I maintain is libertarian—in the philosophical, not political sense—allowing indeterminism and simultaneous free will. This is true, but only to a first approximation, for ultimately I argue that choice emerges from our choice to choose. (Sounds paradoxical? That is not a coincidence.)

19. You could also view this certainty as the ultimate prison, a dungeon built of knowledge.

20. Even here, we do not understand all the details, but sufficiently many to identify key players. R. E. Silversmith and R. E. Bourret, "Throwing the Switch in Bacterial Chemotaxis," *Trends in Microbiology* 7 (1999): 16–22, contains a more detailed account.

21. As usual, I am omitting details in the interest of clarity, such as that more than one CheY-P molecule may have to be bound to the engine and the possibility that the direction switch may follow a stochastic mechanism. For such details, see A. Bren and M. Eisenbach, "How Signals Are Heard during Bacterial Chemotaxis: Protein-Protein Interactions in Sensory Signal Propagation," *Journal of Bacteriology* 182 (2000): 6865–6873.

22. In fact, some of the components involved can be made visible through the attachment of fluorescent dyes or proteins to them, but it is currently impossible to visualize all of the molecules involved simultaneously.

23. Karl von Frisch, *The Dance Language and Orientation of Bees* (Cambridge, MA: Harvard University Press, 1967).

24. The mechanism itself is very interesting. The task is to measure the surface area of a crevice, which is proportional to its volume because of the crevices' geometry. A scout estimates this area by walking across it in an erratic pattern and counting the number of times its path crosses itself. The number of path crossings can be used to estimate the surface area; see, e.g., Eamonn B. Mallon and Nigel R. Franks, "Ants Estimate Area Using Buffon's Needle," *Proceedings of the Royal Society B* 267 (2000): 765–770. Note, however, that this remarkable cognitive feat is different from (and only a prerequisite for) the colony's collective choice.

25. These and many other examples of self-organization can be found in a very readable monograph, Scott Camazine et al., *Self-Organization in Biological Systems* (Princeton, NJ: Princeton University Press, 2001). It is no coincidence that the

foundations of our own choices — rational or not — are similar. Each one of a million nerve cells may fire at an unpredictable moment. Together, however, they carry out the complex and precise calculations required for many rational choices.

26. This raises the question whether any fundamentally random and unpredictable event reflects choice. An affirmative answer would be absurd in the eyes of some and would reduce choice to a trivial and uninteresting notion in the eyes of others. I do not argue for this perspective but merely note that choice and randomness are intimately connected: randomness is the seed of choice. And notice that randomness, built into the foundation of our world, is by no means a trivial phenomenon. It is perhaps one of the world's most fundamental and mysterious properties. Unpredictability and randomness are not only at the heart of choice but at the heart of creation itself.

27. It may also be worth reminding ourselves that twentieth-century physics has eradicated the ancient and naive materialism handed down to us from the Greeks. It has taught us that viewing the world as a mess of tiny bouncing billiard balls is at best a caricature of its complexities. It has also shown that the nature of the smallest — wave or particle — may depend on the questions asked of it, and it has thus cast the immutable nature of matter itself in doubt. Materialism, of course, has clear meaning only when its foundation, the nature of matter, is clear. What becomes of it when the nature of matter itself emerges from a conversation?

28. Another reason why we may not find such organisms relates to standards of scientific inquiry, where inferences about behavior usually require not only single observations but evaluation of the typical or average behavior of an organism or a colony. In other words, counterintuitive observations (bacteria swimming away from food) would often be dismissed post hoc (after theory formation) as "noise" in the experimental data.

Notice also that the question whether a living being could have chosen differently is by no means only difficult to answer for nonhuman organisms. It is also central to the human free will debate. Curiously, though, we tend to deny choice, by this standard, to many nonhuman organisms. The grounds on which to do so are shaky, for several reasons. First, because scientific inquiry involving nonhuman behavior lets us make inferences only from average and repeatable observations, it appears to exclude a priori the possibility of examining choice as carefully as would be necessary. Second, a key but often neglected distinction is that between intentional action, a key ingredient of choice, and voluntary action. Intentional is by no means the same as voluntary. For example, a drug addict may ingest a drug intentionally but not voluntarily, and a prisoner under torture may reveal information intentionally but not voluntarily. For these examples, as well as a lucid discussion of intentionality, voluntary action, and the free will problem, see Jerome A. Shaffer, *Philosophy of Mind* (Englewood Cliffs, NJ: Prentice-Hall, 1968), chap. 5, "The Subject of Consciousness." Third, note that the freedom to choose any one among a set of options is not a necessary ingredient to conceptions of individual autonomy, nor is the existence of a range of options to choose from. This position is forcefully defended by Meir Dan-Cohen, "Conceptions of Choice and Conceptions of Autonomy," *Ethics* 102 (1992): 221–243, who argues that the concept of "willing" is superior to concepts of choices among a set of options in explaining our intuitions of individual autonomy. The main idea behind this view is perhaps best illustrated by the difference of somebody confined to a space (such as a prison cell or a monastery) who chooses to be there (even though he or she can not leave) versus somebody who was coerced to be there.

29. The viral example suggests that this principle applies beyond the living to all entities that self-reproduce because by many definitions viruses are not alive.

30. K. Lorenz, *Behind the Mirror: A Search for a Natural History of Human Knowledge* (London: Methuen, 1977); Karl R. Popper, *Objective Knowledge: An Evolutionary Approach* (Oxford: Clarendon Press, 1972).

31. It is a principle common to many molecular motors that are based on Brownian ratchets. For an overview of some important molecular motors, see M. A. Titus and S. P. Gilbert, "The Diversity of Molecular Motors: An Overview," *Cellular and Molecular Life Sciences* 56 (1999): 181–183.

32. For instance, why is our blood circulation system built like a tree, branching from a massive arterial trunk into smaller and smaller vessels down to capillaries not much wider than blood cells? Because the tree shape allows oxygen to reach all parts of the body. At the same time it allows the heart to pump blood with little resistance, as argued in James H. Brown and Geoffrey B. West, eds., *Scaling in Biology* (New York: Oxford University Press, 2000).

33. I am referring to lithotrophic bacteria that can extract energy from materials such as ferrous iron; see, e.g., Gerard J. Tortora, Berdell R. Funke, and Christine L. Case, *Microbiology: An Introduction* (Menlo Park, CA: Addison Wesley Longman, 1998).

34. You might think that this limited applicability of laws is unique to biology. Not so. A chemical reaction necessary to sustain a living cell may not occur at different temperatures, in acetone instead of water, in the dark, or on a different planet. Chemists can create chemical environments governed not by water but, for example, by ammonium or acetone and environments that know unearthly high pressures or temperatures. In these chemical "worlds," reactions quite different from those in living cells occur. Some of these worlds may have existed in earth's past, on other planets, on the surface of the sun, or somewhere in deep space. Others may not exist anywhere else. And, finally, predictions made by physical laws may also depend on circumstances. Light travels at three hundred thousand meters per second, but only in a perfect vacuum; frictional drag is proportional to velocity, but only at low velocity; and the mass of an object is constant only if its velocity is, too.

A different way of putting all this would be to distinguish between a law and its domain — the Holocene, an aqueous environment, or a perfect vacuum — that is, that which has to be held constant in the *ceteris paribus* condition necessary for the law to apply. It is of course trivially true that the fundamental laws of physics have a broader domain than any fundamental laws of biology, because not all matter is alive.

One issue that I have skirted so far is the distinction between (fundamental) laws and explanations on one hand and predictions on the other. By many conventional accounts, predictions follow from laws. But a careful analysis of this matter, for example, in Nancy Cartwright, *How the Laws of Physics Lie* (Oxford: Oxford University Press, 1983), suggests that this is often not the case, especially for fundamental laws, which do not easily generate predictions. Instead, one often needs a second category of laws, phenomenological laws, from which predictions follow directly (and which represent most of the laws that organisms "discover"). Such phenomenological laws do not always follow from fundamental laws either, as discussed in Cartwright, *How the Laws of Physics Lie;* and in Cartwright, "Fundamentalism Vs the Patchwork of Laws," *Proceedings of the Aristotelian Society* 93 (1994): 279–292.

35. Naturally, these predictions are not easily studied in infants. However, a sizable and fascinating literature exists in this area of cognitive child development. For an example on object unity, see Scott P. Johnson, "Visual Development in Human Infants: Binding Features, Surfaces and Objects," *Visual Cognition* 8 (2001): 565–578. It may seem important for the evolutionary perspective that these predictions are not generally innate—infants learn to make most of them. However, the same argument applies here as for the acquisition of language. Although language itself is not innate, the ability to learn one or several languages surely is—which is where biological evolution comes in. A more complex aspect of the question of innateness is this: much of what seems like acquisition of the abilities I discuss may simply be due to cortical maturation and the gradual development of the ability to extract an amount of information from the environment sufficient to make the relevant predictions.

36. Another case in point is the myriad optical illusions that create distorted sizes, colors, proportions, or spatial relations of objects in our mind's eye. For an accessible introduction, see Donald D. Hoffman, *Visual Intelligence: How We Create What We See* (New York, NY: W. W. Norton, 1998).

37. This becomes even clearer in the world of quantum mechanics, where the observer's question influences a system's answer. Quantum mechanics is a world bizarre beyond our imagination, defying our awkward efforts to dissect it. But it is part of the very world we live in and may thus elucidate flaws in our hidden metaphysical assumptions.

Chapter 7. Purposeful Openness

1. A and B might also be amino acids that will form part of an enzyme. Enzymes make enzymes.

2. Many biochemical reactions that join two molecules A and B are condensation reactions that produce a small-molecule by-product such as water. A and B are

thus strictly not merely joined in these reactions. However, synthesis reactions or direct combination reactions that do not produce any by-products also exist.

3. J. Piatigorsky and G. J. Wistow, "Enzyme/Crystallins: Gene Sharing as an Evolutionary Strategy," *Cell* 57 (1989): 197–199; Stanislav I. Tomarev and Joram Piatigorsky, "Lens Crystallins of Invertebrates: Diversity and Recruitment from Detoxification Enzymes and Novel Proteins," *European Journal of Biochemistry* 235 (1996): 449–465.

4. Cells face an additional danger that houses do not face. Specifically, a hypo-osmotic environment may lead to osmotic swelling caused by water uptake into the cell. The cytoskeleton may also be involved in maintaining a cell's integrity in such an environment.

5. For an elaboration of this perspective that highlights a construction element important in church architecture, see Stephen Jay Gould and Richard C. Lewontin, "Spandrels of San Marco and the Panglossian Paradigm: A Critique of the Adaptionist Program," *Proceedings of the Royal Society B* 205 (1979): 581–598.

6. Even more exotic purposes have also been proposed for these proteins, such as that they harbor the seeds of the human mind. Roger Penrose elaborates on the role of microtubules in neural computation in Penrose et al., *The Large, the Small and the Human Mind,* ed. Malcolm Longair (Cambridge: Cambridge University Press, 1997).

7. Gabrielle Kardon, "Evidence from the Fossil Record of an Antipredatory Exaptation: Conchiolin Layers in Corbulid Valves," *Evolution* 52 (1998): 68–79, addresses questions about the evolutionary origins of conchiolin. The following example raises similar questions, including the question whether some features of organisms might, at least originally, serve no purpose at all. Following a suggestion by Stephen Jay Gould and Elisabeth S. Vrba, such features are often also called exaptations, a term introduced in Gould and Vrba, "Exaptation: A Missing Term in the Science of Form," *Paleobiology* 8 (1982): 4–15. Whether exaptations exist has been the purpose of lively debates; see Stephen Jay Gould, "The Exaptive Excellence of Spandrels as a Term and Prototype," *Proceedings of the National Academy of Sciences* 94 (1997): 10750–10755. However, this question is peripheral for my purpose, which is to point to the enormous creativity of the living in creating purpose.

8. W. Scott Armbruster, "Exaptations Link Evolution of Plant-Herbivory and Plant-Pollinator Interactions: A Phylogenetic Inquiry," *Ecology* 78 (1997): 1661–1672, discusses this example in detail.

9. For examples, see P. Nicholls, "Introduction: The Biology of the Water Molecule," *Cellular and Molecular Life Sciences* 57 (2000): 987–992; and Y. Pocker, "Water in Enzyme Reactions: Biophysical Aspects of Hydration-Dehydration Processes," *Cellular and Molecular Life Sciences* 57 (2000): 1008–1017.

10. What about the smallest objects we know of, elementary particles? They are supposed to be simple, these smallest conceivable billiard balls that bounce endlessly into one another. But each has a series of properties, some familiar—like

mass — others strange — like spin. Different combinations of these properties results in different particles, of which physicists have already named several dozen. And who is to say that the number of "elementary" particles we recognize might not grow, perhaps indefinitely? For the more authoritative voice of a physicist on this topic, see Roger S. Jones, *Physics as Metaphor* (Minneapolis: University of Minnesota Press, 1982), 116–117. And even if they will turn out to be finite in number, these particles have a unique feature I already mentioned several times. Quantum laws dictate that it is utterly unpredictable whether any two such particles will encounter each other. Thus, although two particles' properties may determine an encounter's outcome, we cannot know whether the encounter ever occurs.

11. D. Jean, K. Ewan, and P. Gruss, "Molecular Regulators Involved in Vertebrate Eye Development," *Mechanisms of Development* 76 (1998): 3–18.

12. Scott F. Gilbert, *Developmental Biology,* 5th ed. (Sunderland, MA: Sinauer, 1997), 41.

13. Pierre Teilhard de Chardin, *Christianity and Evolution,* trans. René Hague (New York: Harcourt Brace Jovanovich, 1971).

14. See, e.g., Francisco J. Ayala, "Darwin's Devolution: Design without Designer," in *Evolutionary and Molecular Biology: Scientific Perspectives on Divine Actions,* ed. Robert John Russell, William R. Stoeger, SJ, and Francisco J. Ayala (Notre Dame, IN: University of Notre Dame Press, 1998); Hugh Lehman, "Functional Explanations in Biology," *Philosophy of Science* 32 (1965): 1–20; and Michael E. Ruse, "Function Statements in Biology," *Philosophy of Science* 38 (1971): 525–528.

15. For a delightful if anecdotal analysis of human innovation, see Tom Kelley, *The Art of Innovation:* Lessons in Creativity from IDEO, America's Leading Design Firm (New York: Currency/Doubleday, 2001). A more scholarly treatise is Robert J. Sternberg, ed., *Handbook of Creativity* (Cambridge: Cambridge University Press, 1999).

16. Consider that the difference between the random mutations leading to new biological molecules and the process by which humans create innovations is also a matter of degree, even though the material substrate of innovation may be quite different.

17. Are there, then, no fundamental differences between the power of humans and other animals to generate purpose and meaning? I see only one. Humans (and perhaps a few other animals) know about themselves. Humans can endow themselves with a place in the world. They can create purpose and shape not only other things but also themselves. Where are the limits of this ability to shape not only other but also self, not only matter but also meaning and mind, and thus to participate in the creation of a world?

18. This question is very similar to the unsolved question "What is meaning?" alluded to earlier.

19. A lucid exposition of this proposal is given by L. Wright, "Functions," *Philosophical Review* 82 (1973): 139–168.

20. A collection of articles discussing many facets of this issue can be found in

David Hull and Michael Ruse, *The Philosophy of Biology* (New York: Oxford University Press, 1997).

21. You might even choose to view function as the — inexhaustible — potential of an object to enter relations.

22. Who is having eyes bad for?

23. An important qualification is that some cancers are caused by infectious agents and may thus be either by-products of the infection or even serve a purpose for the infectious agent.

24. If the sophisticated design features of complex organs like eyes or of complex cellular structures like the cytoskeleton are evidence of an "intelligent" designer, then the barbed stings of bees or the nihilism of tumors might be construed as evidence of a "dumb" designer.

25. Strictly speaking, there are beneficiaries of oncogenesis. These are the cells that had stopped dividing, that were only to serve the greater good of the organism, and that were destined to die with it. When a tumor forms, these cells start dividing and produce offspring again. They are of course granted only a short reprieve. But then, again, any lineage of organisms or cells is doomed to extinction. It is just a matter of time.

Chapter 8. Choice and the Natural Sciences

1. By following an anecdotal style here, I do not mean to ignore the vast body of relevant work that exists in the philosophy of science. I just highlight observations I consider particular important, to avoid getting bogged down in a discussion of the many "-isms" that are accessible only to the academic philosopher. You can find a representative selection of other material in the contemporary academic philosophy of science in David Papineau, ed., *The Philosophy of Science* (Oxford: Oxford University Press, 1996).

2. One might argue that there are two exceptions to the primacy of explanation. First, in statistical forecasting, such as in the elaborate time-series forecasting models used in finance, the goal is to predict the future values of some variable(s) of interest from past values without having a detailed "mechanistic" model of their interaction. However, even such statistical models contain implicit assumptions about causally relevant (explanatory) variables, and a specific mathematical apparatus to posit a relation of the independent variables to the variables whose behavior is to be predicted. Second, phenomenological laws predict mathematical relations of some observables (for example, the expansion of a metal rod as temperature increases or the half-life of different radionuclides). Although no causal (explanatory) relation of the relevant variables is obvious from the phenomenological law itself, the law usually emerges from an underlying more "fundamental" law that posits such causal relations. The relation of fundamental and phenomenological laws is by no means simple, as Nancy Cartwright has pointed out in *How the Laws of Physics Lie* (Oxford: Oxford University Press, 1983), chap. 6.

3. The question "What is an explanation?" appears to defy a rigorous answer. The problems are so fundamental that brief introductions are even available in encyclopedias of philosophy. Some such encyclopedias are available online and provide further extensive references; see, e.g., the Internet Encyclopedia of Philosophy at http://www.utm.edu/research/iep.

4. Sir Alexander Fleming's original discovery was published in "On the Antibacterial Actions of Cultures of a Penicillium, with Special Reference to Their Use in the Isolation of B. Influenzae," *British Journal of Experimental Pathology* 10 (1929): 235–236. Fleming ended his own research on penicillin a few years later. His work was continued by others, most notably Ernst Chain and collaborators, who first established the therapeutic usefulness of penicillin; see Chain et al., "Penicillin as a Chemotherapeutic Agent," *Lancet* 2 (1940).

5. For the sake of accuracy, this is only one of the colors Fleming used in describing the fungus *P. punctatus*. Fleming, "On the Antibacterial Actions of Cultures of a Penicillium." Any other color could serve to make the same point.

6. Note another description built around a slew of interpretations: visible light may be composed of light of many wavelengths.

7. Think of the countless visual illusions we are subject to. For a popular introduction, see Donald D. Hoffman, *Visual Intelligence: How We Create What We See* (New York: W. W. Norton, 1998).

8. The best point to delve into this subject is a text on cognitive psychology; see, e.g., John R. Anderson, *Cognitive Psychology and Its Implications,* 3rd ed. (New York: W. H. Freeman, 1990).

9. Hence the need to hold the conversations of cognitive psychology that are dedicated to showing these interpretations as what they are. These conversations, of course, are but another departure point into an endless maze.

10. I do not argue that they are identical in every respect, just in those most important for our purpose. One could draw many other distinctions here, including those between declarative, tacit, and procedural knowledge often made in cognitive science; see Neil A. Stillings et al., *Cognitive Science: An Introduction* (Cambridge, MA: MIT Press, 1987).

11. I am deliberately using the words "interpretation" and "explanation" interchangeably.

12. Although the plant uses nothing like our nervous system to arrive at this interpretation, molecules nevertheless communicate to create the interpretation, just like in our brain. If you wished to pinpoint the part of the plant responsible for this interpretation, you would get entangled in the same kind of problem I discussed earlier for swimming bacteria.

13. This perspective on explanations is nothing but a generalization of Hume's characterization of causes as associations among objects or events. Events or objects become signs in the sense I discussed earlier because they can stand for the causes behind themselves. This is how the notion of "meaning" comes in here.

14. Cognitive psychology distinguishes here between automatic and controlled processes.

15. The Copernican revolution is at the core of a study that spawned many other attempts to reconstruct the history of scientific theories and how they succeed each other: Thomas S. Kuhn, *The Structure of Scientific Revolutions* (Chicago: University of Chicago Press, 1962). Observations unaccounted for by an existing theory are critical for this succession. Kuhn stresses that successive theories may be meaning-incommensurable — that is, their terms cannot be mapped onto each other. Although this is likely to hold for some theories, it may not hold for others, such as the thermal contraction and continental drift theories; see Naomi Oreskes, *The Rejection of Continental Drift: Theory and Method in American Earth Science* (New York: Oxford University Press, 1998).

16. This is another subject on which a large amount of literature exists. A historical account of the development of special relativity is given in Arthur I. Miller, *Albert Einstein's Special Theory of Relativity: Emergence (1905) and Early Interpretation (1905–1911)* (Reading, MA: Addison-Wesley, 1998).

17. Strictly speaking, the hypothesis tested in the famous Michelson-Morley experiment was that two light beams traveling at an angle relative to each other would propagate at different speeds. This would be equivalent to measuring the speed of the same light beam at different times during earth's rotation, except that it is more practical, because the relative velocities of the two beams can be compared by letting them interfere directly with each other.

18. Unlike many other scientific explanations, this explanation has not become automatic or nearly so — perhaps because it was so hard to choose in the first place, perhaps because we do not experience time differently depending on how fast we move, or perhaps because we have difficulties relating to the explained observations in everyday terms.

19. Not all of Einstein's choices have been equally successful. Years after he first proposed his principle of relativity, another group of physicists, including Max Planck, Werner Heisenberg, and Erwin Schrödinger, developed quantum theory, a perspective that explained other problematic "details" of classical physics. Quantum theory is another huge success story in physics. (Einstein also made an early important contribution to it, in a 1905 paper suggesting that light might have particulate nature.) Countless experiments confirmed many of quantum theory's predictions, and physicists work with it to this day.

One of quantum theory's most important consequences is that unpredictability and randomness is fundamental to physics. At least this is the most widely accepted interpretation of the mathematical results of the theory, known as the Copenhagen interpretation; see James T. Cushing, *Philosophical Concepts in Physics: The Historical Relation between Philosophy and Scientific Theories* (Cambridge: Cambridge University Press, 1998). Einstein did not think this essential randomness was the last word. He was unwilling to choose this perspective. As he famously stated in a letter to Niels Bohr, "God does not play dice." He believed that the Copenhagen interpretation's view on randomness was a temporary fix (similar to the ether that Einstein himself had done away with). It would be abandoned at some point, as better perspectives became available. And he remained committed to this choice. Most contemporary physicists would agree that he was wrong on this one. Randomness may be here to stay.

20. You can find a detailed, albeit technical account of this perspective in Oreskes, *Rejection of Continental Drift*.

21. The German original is Alfred L. Wegener, *Die Entstehung der Kontinente und Ozeane* (Braunschweig: Friedrich Viewig, 1915). It appeared in translation as Wegener, *The Origins of Continents and Oceans* (New York: Dover, 1966).

22. For a historical account, see Ernst Mayr, *The Growth of Biological Thought: Diversity, Evolution, and Inheritance* (Cambridge, MA: Belknap Press of Harvard University Press, 1982).

23. Here are a few of the many additional points that are worth making about Wegener's continental drift theory (Oreskes, *Rejection of Continental Drift*), Darwin's theory of evolution by natural selection (Mayr, *Growth of Biological Thought*), and Einstein's special relativity (Miller, *Einstein's Special Theory of Relativity*), as well as about many other theories. First, the choices at issue are not simple. They require the chooser to distill a huge body of information and to use the distillate ingeniously to reorganize the whole body of information. Second, contemporaries of all three individuals, such as Wallace in the case of Darwin or Lorentz in the case of Einstein, had made similar choices. The reasons why these choices did not get as much attention are varied. Third, historians have made serious efforts to try to understand why novel perspectives are not accepted. The most important message from their studies is that the reasons have less to do with evidence presented for a theory and much more with accepted wisdom or accepted practice of doing science. Fourth, it is usually possible to provide "fixes" to most problems a theory faces. It is not always easy to see why a new and radically different choice eventually prevails, despite existing fixes to an old theory. A common argument is that the new theories are simpler or more elegant, a mostly aesthetic criterion that is hard to make precise.

24. An account of several biologists whose choices went against accepted wisdom can be found in Oren Harman and Michael R. Dietrich, eds., *Rebels, Mavericks, and Heretics in Biology* (New Haven and London: Yale University Press, 2008). For a thorough analysis of the sociology of scientific knowledge, see Barry Barnes, David Bloor, and John Henry, *Scientific Knowledge: A Sociological Analysis* (Chicago: University of Chicago Press, 1996), esp. chap. 2.

25. The philosophy and history of science has largely been one of the history of theory. There are notable exceptions, studies that try to delineate in great detail how observations and experiments are made and how difficult the judgments necessary to interpret an observation or experiment really are; see Cushing, *Philosophical Concepts in Physics;* Peter Galison, *How Experiments End* (Chicago: Univer-

sity of Chicago Press, 1987); and Andrew Pickering, *Constructing Quarks: A Sociological History of Particle Physics* (Edinburgh: Edinburgh University Press, 1984).

26. Even so, special conversations, such as those of cognitive psychology, are often necessary to show us our interpretations as what they are.

27. Nowadays, this ultimate reflection is very complicated. A lifetime would not be enough to absorb it. In contrast, a few thousand years ago, one person could acquire the body of human knowledge. The ultimate interpretation has also become much more abstract since then. It now fills hundreds of thousands of volumes and uses such terms as elementary particles, spins, quarks, DNA, magnetic moments, proteins, and so on. We have no direct experience of many of these concepts. (But curiously, any textbook explaining such terms to science students uses metaphors like billiard balls, water waves, pendulums, and guitar strings. Perhaps we have not gone all that far after all?)

Chapter 9. The Limits to Knowledge

1. Even this certainty may be too much. For instance, Carol E. Cleland, "Methodological and Epistemic Differences between Historical Science and Experimental Science," *Philosophy of Science* 69 (2002): 474–496, points out that an experimental test of a hypothesis involves testing not only of the hypothesis itself but implicitly also of multiple auxiliary assumptions. Modifications of these assumptions may lead to different experimental outcomes. This means that some refutations of a hypothesis may be "false negatives," caused by failure of particular auxiliary assumptions.

2. As an aside, notice that to emphasize questions that allow a negative answer distracts from the essential human act in creating these questions. The world provides no recipe for what questions to ask and for how to build new interpretations on the answers to these questions. It merely crushes old interpretations. But the creation of new interpretations, new choices, and new questions is a creative act without equal. It is quintessentially human but also the most neglected aspect of scientific conversations, perhaps because we know little about it.

3. Most scientists would concede that it is important to be able to attack a theory's soft underbelly. And they might agree that there is no rigorous approach to verifying whether a theory is true. But their agreement contains a curious double standard. Were they true to their word, they would busily dismantle their theories, because only a theory's destruction creates the empty space that new theories can fill. Instead, most of us do the exact opposite. They cling to their theories, defend them fiercely against any onslaught, and attack instead their opponents. Science is rife with controversies that testify to this double standard. Once again, our biological past may not be easily shaken.

4. This idea is by no means new, as I mentioned earlier. Its most visible exponent is evolutionary epistemology, a branch of epistemology whose origins date back to the work of Darwin. The term was coined by Donald Campbell; see

Campbell, "Evolutionary Epistemology," in *The Philosophy of Karl Popper*, ed. Paul Arthur Schilpp (La Salle, IL: Open Court, 1974). An exhaustive reading list can be found in Gary A. Cziko and Donald T. Campbell, "Comprehensive Evolutionary Epistemology Bibliography," *Journal of Social and Biological Sciences* 13 (1990): 41–81. This list is also available in updated form online (http://faculty.ed.uiuc.edu/g-cziko/stb).

5. Larry Laudan has termed this position convergent epistemological realism. He refutes it in an essay much more rigorous than my brief analysis of a metaphor; see Laudan, "A Confutation of Convergent Realism," *Philosophy of Science* 48 (1981): 19–49. Laudan presents a number of arguments against convergent epistemological realism, some of which I will summarize briefly. First, he asks, are theories whose central terms refer (to some entities in a reality independent of us) inevitably successful? The answer is no. For example, atomistic theories in chemistry were not successful up to the nineteenth century although (we think) their key terms — atoms — refer. Second and conversely, are successful theories those whose terms refer? No. There are plenty of past successful theories — most notably ether theories in chemistry and physics — whose terms (we now think) do not refer. Thus, the link between reference as a measure of truth and success of a theory is tenuous. Laudan then examines the following statement (which is weaker because it does not require reference): "If a theory is approximately true, it will be explanatorily successful" and its converse. He presents a plethora of examples of past explanatorily successful theories — including the caloric theory of heat and catastrophist geology — that we hold now not to be true. He also argues that there is no clearly applicable, operationally useful notion of "approximate truth" that could be applied to resolve this question. He finally dismantles a prominent argument of convergent realists — namely, that later theories retain central terms of earlier theories and that earlier theories can often be regarded as limiting cases of later theories, which implies some notion of progress in these theories. Counterexamples include Copernican astronomy and Newtonian physics; see Laudan, "Confutation of Convergent Realism," 127–128. The latter point has perhaps been made most lucidly by Thomas S. Kuhn, *The Structure of Scientific Revolutions* (Chicago: University of Chicago Press, 1962).

6. There are multiple theoretical attempts to compare the truth content of theories, some of which are reviewed in Laudan, "Confutation of Convergent Realism." Curiously, they have never been applied — perhaps because it is impossible to do so? — to compare the truth content of *actual* scientific theories.

7. To view laws of nature as metaphors exemplifies a philosophical position that views essentially all basic concepts we use to describe the world, including central philosophical notions, such as cause and effect, as metaphorical. Perhaps the most forceful advocates of this position in cognitive science are George Lakoff and Mark Johnson, who have presented it in an account for a general readership (Lakoff and Johnson, *Metaphors We Live By* [Chicago: University of Chicago Press, 1980]), in

an exploration of metaphors used in mathematics (Lakoff and Rafael E. Núñez, *Where Mathematics Comes From: How the Embodied Mind Brings Mathematics into Being* [New York: Basic Books, 2000]), and in a philosophical treatise (Lakoff and Johnson, *Philosophy in the Flesh: The Embodied Mind and Its Challenge to Western Thought* [New York: Basic Books, 1999]). Their position ultimately rests on the realization that we cannot separate our cognitive abilities from the neural and physical basis needed to exert them. You will find their analysis of metaphors much more thorough than the anecdotal account I give here. That any metaphor has to fail is not surprising, as metaphors are based on similarity to their representamen (on isomorphism, to employ the term used in Douglas R. Hofstadter, *Gödel, Escher, Bach: An Eternal Golden Braid* [New York: Basic Books, 1979]). Only the representamen is identical to itself, so any metaphor must have features that do not map onto its representamen.

8. Recall, though, that this is no guarantee for receiving the same answer every time you *could* ask.

9. Where to draw the line between what matters and what does not for understanding a natural phenomenon may require difficult choices, choices that may be obvious only in hindsight. Sometime it may need sheer luck. For instance, *Penicillium* produces much penicillin only when mildly starved. Well fed, it does not assassinate its competitors — it simply has no need to. Similarly, it matters how acidic the environment of the fungus is, for acid destroys penicillin. To work only with starved *Penicillium* in a nonacidic environment is not a priori an obvious choice.

10. Only in limited areas of science, such as computer science, does the notion "complexity" have a precisely defined meaning. In others, we have to contend ourselves for now with enumerating features of complex systems such as the ones I describe here.

11. As I said earlier, this holds even for some systems where only few factors matter, like chaotic chemical reaction systems. Such "simple" yet unpredictable systems are not subject to the impossibility of repeatable questioning. They show that sometimes even being able to ask approximately the same question twice is not good enough.

12. Statistics is arguably a successful approach to the unique. However, it does not explain the individual event, which is all that matters to the individual.

13. I emphasized elsewhere that our uncertainty about the future need not be absolute. For if uncertainty rules the world, so does uncertainty about this uncertainty. Science provides a good example of this tension: neither are laws of nature final, nor is our defeat in finding them. Take the ideal gas law, where a simple principle connecting pressure, temperature, and volume emerges from the erratic interactions of innumerable molecules. This example, however, is different from the examples in the main text, because the system at issue is assumed to be in equilibrium.

14. Geometry as we know it — Euclidean geometry — is only one possible kind of geometry. Geometries in curved spaces, such as Riemannian geometry, have proven equally or more important in explaining physical phenomena.

15. Mathematicians are divided among themselves as to whether they are inventors or discoverers; see Allen L. Hammond, "Mathematics: Our Invisible Culture," in *Mathematics Today: Twelve Informal Essays,* ed. Lynn Arthur Steen (New York: Springer, 1978).

16. Gödel's first incompleteness theorem was published in K. Gödel, "Über formal unentscheidbare Sätze der Principia Mathematica und verwandter Systeme, I," *Monatshefte für Mathematik und Physik* 38 (1931): 173–198. A less technical account of relevant material — and paradoxes in general — is contained in the excellent book by Hofstadter, *Gödel, Escher, Bach,* as well as in many informal essays by mathematicians, e.g., Martin Davis, "What Is a Computation?" in *Mathematics Today: Twelve Informal Essays,* ed. Lynn Arthur Steen (New York: Springer, 1978).

17. This is the subject of Gödel's second incompleteness theorem.

18. Examples include the absence of an algorithm for deciding whether a given Diophantine equation, a polynomial with integer coefficients, has an integer solution and, more important, the mathematical theory of randomness, where the proposition that any given real number is a random number is unprovable; see Gregory J. Chaitin, *Algorithmic Information Theory* (Cambridge: Cambridge University Press, 1987). For nontechnical treatments, see Chaitin, "Randomness and Mathematical Proof," *Scientific American* 232 (1975): 47–52; Chaitin, "A Century of Controversy over the Foundations of Mathematics," *Complexity* 5 (2000): 12–21; and Chaitin and C. S. Calude, "Randomness Everywhere," *Nature* 400 (1999): 319–320.

19. A. Turing, "On Computable Numbers with an Application to the Entscheidungsproblem," *Proceedings of the London Mathematical Society, Series 2* 42 (1936–1937): 230–265.

20. In addition, this machine uses a finite number of symbols and is capable of adopting a finite number of computational states.

21. Although the theory of computation is dominated by concepts built on discrete computation, some mathematicians have aimed at a more expansive view of computation, e.g., J. F. Traub and E. W. Packel, "Information-based Complexity," *Nature* 327 (1987): 29–33; J. F. Traub and H. Woźniakowski, "Breaking Intractability," *Scientific American* 270 (1994): 102–107; and Hava T. Siegelmann, "Computation beyond the Turing Limit," *Science* 268 (1995): 632–637.

22. The result of a computation is then not necessarily deterministic — just think of three gyrating planets as computers — although computers are traditionally viewed as deterministic devices. But that position may need reconsidering, for even some man-made computers are no longer deterministic — if not in principle, then in practice. Chess computers already beat world champions. And thus, no human could understand what is going on inside that machine during a match, except of course in hindsight, after weeks of painstaking analysis. And for com-

puters inside living organisms the situation is worse. A bird may arrive reliably at its nest. But the chatter of nerve cells necessary for this computation would appear erratic to even the most sophisticated instrument available today. Even worse is the swimming bacterium. Not only are we unable to predict how each of its parts contributes to the computation. The computation's outcome, the decision whether to change direction at any moment, is also unpredictable.

23. Examples include the antique antikethyra and astrolabe (likely ancestors of the planetarium), analog computers using hydrostatic principles to solve algebraic equations, mechanical analog computers to solve differential equations, and electronic analog computers used in ballistic calculations during World War II. A detailed account of analog computers is provided by Georges Ifrah, *The Universal History of Computing: From the Abacus to the Quantum Computer,* trans. E. F. Harding (New York: Wiley, 2001).

24. In the 1920's, before Turing's work, the mathematician David Hilbert pointed to the problem of finding a computing procedure to test whether a given theorem is true as one of the most important outstanding problems of mathematics. Turing's results showed that this problem is unsolvable.

25. The Möbius strip's value as an analogy to the mathematical problems I mentioned has been highlighted by many others, most notably in Hofstadter, *Gödel, Escher, Bach*.

26. One of the most remarkable of these is Schrödinger's cat paradox, from E. Schrödinger, "Die gegenwärtige Situation in der Quantenmechanik," *Naturwissenschaften* 48 (1935): 807–812.

27. This does not hold so much for Gödel's paradox, but it is obvious for others, such as Russell's paradox of sets that do not contain themselves; see W. V. Quine, *The Ways of Paradox, and Other Essays* (Cambridge, MA: Harvard University Press, 1976).

28. Could statements like these be proven wrong? Yes, if we all reached agreement on how to resolve the philosophical tensions — between self and other, whole and part, and so on — that occupy much of this book. But if anything, we may be moving farther away from such resolution.

Chapter 10. The Power and Burden of Freedom

1. Perhaps the design of von Neumann computers reflects tellingly on our neglect of the relation between matter and mind.

2. K. R. Popper, *The Open Society and Its Enemies,* 2 vols. (Princeton, NJ: Princeton University Press, 1962), discusses this and other paradoxes in chap. 7 and in notes 4–6 of the same chapter. He correctly points out that every form of government built on a sovereign — whether the People, the "wisest," or the "best" individual — harbors this paradox, because the wisest could rule that not he or she but the People should govern, and so on. The dangers that come with this paradox can be reduced — but not eliminated — through institutional govern-

ments, in which a balance of power among several institutions prevents sovereignty by any one of them. The history of U.S. government institutions such as the Supreme Court testifies to the incessant power struggles among these institutions; see, e.g., Peter Irons, *A People's History of the Supreme Court* (New York: Viking, 1999). Governments involving checks and balances on power are not free from paradoxes. Most of these derive from the fact that no "rational" method may exist to infer a collective preference from preferences of multiple individuals or institutions. Similar problems exist in choices individuals have to make about themselves. An account of this class of problems can be found in Thomas C. Schelling, *Choice and Consequence* (Cambridge, MA: Harvard University Press, 1984).

3. Douglas R. Hofstadter, *Gödel, Escher, Bach: An Eternal Golden Braid* (New York: Basic Books, 1979), chap. 7, n. 4.

4. I owe this example to Hofstadter, *Gödel, Escher, Bach,* chap. 20.

5. Similar tensions in the U.S. Constitution are explored much more exhaustively by Daniel N. Hoffman, *Our Elusive Constitution: Silences, Paradoxes, Priorities* (Albany: State University of New York Press, 1997). Irons, *People's History of the Supreme Court,* discusses the many political tensions associated with the court's decisions back to the court's origins.

6. Schelling, *Choice and Consequence,* chap. 4.

7. Albert Camus, *The Myth of Sisyphus, and Other Essays,* trans. Justin O'Brien (New York: Vintage Books, 1955).

8. Some might argue that an appropriate framework is already in place, for example in well-known notions of computer science such as Shannon information and entropy. However, the existing theory of information and communication presupposes an agent who is able to detect meaning in a message. Shannon information, for example, provides a purely syntactic and statistical measure of information. One of its limitations is that it tells us little about communication from a semantic meaning-centered point of view.

9. Scientists, especially physicists, have often expressed surprise at the enormous success of their theories. Their mathematical apparatus accurately describes stars millions of light-years away, and a subatomic world minute beyond comprehension. Such power would indeed be surprising if humans were as alien to the world as we had invented ourselves to be.

10. What is the blind spot of solipsism?

11. A smattering of other ethical ideas and their shortcomings is contained in a refreshing and readable little book by Simon Blackburn; see Blackburn, *Being Good: An Introduction to Ethics* (Oxford: Oxford University Press, 2001).

12. Translation from the French by the author, with help from Claude Senninger and Carolyn Simmons.

RNA (ribonucleic acid), 64, 217(n6)
Russell, Bertrand, 186, 249(n27)

salamanders, 160
Schelling, Thomas, 137
science as conversation: "approaching reality" metaphor, 181–182; biology, 27; future of, 191; generally, 26; impossibility of proving theories, 179–181, 245(nn1, 3); limitations of science, 198; logocentric perspective and, 200, 250(nn8, 9); nature as partner, 200–201; "no" as answer, 179–181, 245(nn1, 2); partners changed by, 28–29; penicillin discovery, 166–168; primacy of explanation, 165–166, 169, 241(n2); repeated questioning, 182–184, 247(nn8, 9); revolutionary choices, 170–178; social sciences, 26–27; theories defended, 245(n4); types of questions asked, 180–181, 245(nn2, 3). See also mathematics; physics; quantum physics; scientific explanation(s); scientists; and specific individuals, laws, and theories
scientific explanation(s): complexity, 245(n27); defense of, 177–178; defining, 242(n3); description and, 166, 168; as metaphors, 182, 246–247(n7); primacy of, 165–166, 169, 241(n2); sensory experience and, 169; success/failure of, 180–182. See also science as conversation; and specific fields, theories, and scientists
scientists: measuring complexity, 112, 228(n46); study of parts and wholes, 193–194; subject influenced by, 69, 73–74, 238(n37); subject's atypical behavior often dismissed by, 236(n28); surprised by theories' success, 250(n9); synthetic chemicals created, 158. See also science as conversation; and specific scientists
segregation, racial, 137–138
selective breeding, 175–176
self and other: benefits of worldview, 51; distinguishing between, 42–43, 195; universal relatedness/separation, 3, 59–

61, 149, 195. See also altruism; competition; cooperation; parasites; selfishness
self-defense, 157–158
"selfish gene" theory, 42–45, 213(n17–18)
selfishness, 2, 51–53. See also altruism
senses, human, 168–169. See also eyes
separation, universal, 3, 59–61, 149, 195
shellfish, 157
signals: effects as signs for, 30; insect parasites, 9–12, 207–208(n15–17); in molecular communication, 18–19, 22–25, 209(n30–31); plants' chemical signals, 10–11, 20–21, 207–208(n16). See also communication; molecular communication
signs: defined, 8; and meaning, 8–9, 31, 206–207(n7–11), 211(nn48, 49). See also communication
single-celled organisms, 45–48, 50–51. See also bacteria; viruses; Volvocaceae
slime molds, 78–79, 81–82
Smith, John Maynard, 112
snakes, 107, 115
social sciences, 26–27, 210(n40)
Socrates, 4, 198
solar energy. See plants
species, biological: classification, 84–87, 149; genetic drift and species formation, 85–87, 221(nn30–32); reproduction and, 35–36; species boundaries, 83–84, 87–89, 149. See also individual organisms
Sphenodon, 109
squid, 16
statistical forecasting, 241(n2)
stock market, 97–98
strawberries, 82
structure vs. function, 151–152
suicide, 231(n11). See also cells: cell death
symbiotic relationships, 49–51, 214–215(n32)
symbols. See signs
Szent-Györgi, Albert, 176

tadpoles, 122
Talmud, 59
termites, 44, 50–51, 82
thermal contraction theory, 172–175, 177